Morse Cod

e

by L. Peter Carron, Jr., W3DKV

ABOUT THE COVER

Rick Palm, K1CE, uses one of the well-equipped visitors' positions at W1AW to exercise his CW skills. (*Photo by Kirk Kleinschmidt, NTØZ.)*

Published by the
AMERICAN RADIO RELAY LEAGUE
225 Main Street
Newington, CT 06111

Printed in USA

Second Edition
Third Printing, 1996

ISBN: 0-87259-035-6

PREFACE

The Morse code is steeped in tradition, not only in Amateur Radio but in other services as well. The sight of a radio operator sending a message in code generates a certain intrigue that makes the mind wander to thoughts of ships in distress, signals-in-the-night and faint transmissions from distant lands. Although some of the thoughts it conjures are of times gone by, the code is as useful today as it was the day of its invention—more so in many ways. Portions of our high-tech, modern-day military make extensive use of the code. Whether you are a newcomer to Amateur Radio or you have been licensed for 50 years, you owe it to yourself to become proficient in this most valuable means of communication.

ACKNOWLEDGMENTS

Many individuals have contributed to the material in this book, and the author wishes to express his appreciation to all. Wherever possible credit has been given to the appropriate individual by notations within the text. Special thanks are conveyed to Louise Ramsey Moreau, W3WRE, Tony Smith, G4FAI and Bill Fisher, W2OC.

CONTENTS

CHAPTER ONE

Why the Code?

This is the age of the radio, television, satellites, fiber optics and computers. Why would anyone use the Morse code?

There are many reasons.

First, the code can be used when other modes are unavailable (which is often the case), or when they are available, but not in working order.

In addition, the code has distinct advantages of its own. The following are some of its major benefits.

It is the most widely recognized means of signaling in the world.

It is the only code understood by both man and machine.

It is the only code allowed on *all* amateur frequencies.

It is a method of communication that can "get through" and be understood when all others fail. For example, a barely audible code signal can still be deciphered, whereas the same strength voice transmission will be unintelligible.

It is often the only type of signal that can be read through enemy jamming during military operations. For this reason, the 5th Combat Communications Group at Warner Robins Air Force Base in Georgia received orders from higher headquarters to reinstate training recruits in its use. It had been dropped years earlier, replaced by new, more exotic communication modes.

Some types of communication efforts rely almost exclusively on

Morse code because of its ability to succeed where other means fail. This is true in amateur EME (earth-moon-earth) contacts. In this type of activity, amateurs send messages to each other by bouncing signals off the moon. These transmissions reflect back to earth, making possible two-way contact with stations thousands of miles away. Signals reflected from the moon are usually so weak that Morse code is the *only* mode of transmission that can get through. Admittedly, this facet of the hobby is not for the weak-hearted, for it requires sophisticated transmitting and receiving gear. But you never know where your interest in ham radio will lead. Someday, EME may be your thing—so be prepared.

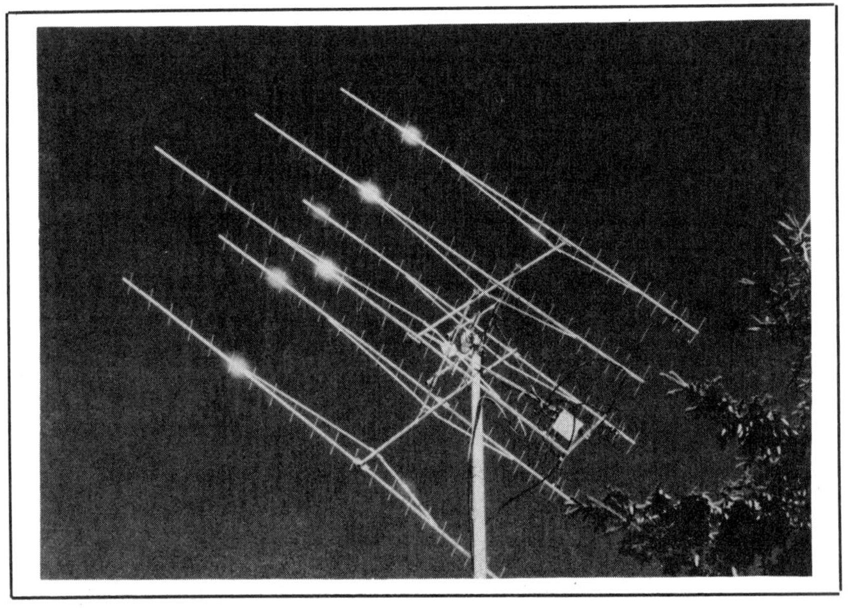

Morse code and big antennas are the order of the day when you're reaching for the moon. (photo courtesy Tim Pettis, KL7WE)

The code also has advantages in that it can be sent by any number of means when voice communication is impossible or when a method of sending messages by other than radio is essential—such as during military radio blackouts. At these times, other means, such as Navy signal lights, are preferred. Flags, mirrors, any means of creating a sound or any means of movement can be used. Code can be transmitted by any method that can be varied into two states such as ON or

OFF, UP or DOWN, LEFT or RIGHT, HIGH or LOW, BLACK or WHITE, BRIGHT or DIM. Its method of transmission might be detected by sight, sound or touch. Code has been sent by automobile horns, railroad whistles, flashlights, tapping knuckles, flashing mirrors and many other means. The possibilities are endless.

Perhaps the most dramatic story in recent history of non-radio transmission of code came out of the Vietnam War. During a forced television interview, Prisoner-of-War Jeremiah A. Denton Jr, blinked

his eyes in Morse code, spelling out the word T-O-R-T-U-R-E in an attempt to notify the outside world of his plight. When interviewed, Denton, later to become a US Senator from Alabama, commented: "Had I not known the Morse code, I would have been denied the one viable option of communications open to me while a prisoner of war."

And there are more advantages to the code. For example, it can be enciphered to prevent discernment of a message by third parties when secrecy is mandatory.

By using an internationally understood set of abbreviations known as Q signals, it is possible to carry on a conversation in Morse between individuals of different nations who do not understand each other's native tongues.

It is cheaper and easier to manufacture a code-only transmitter than any other type. When all the modulators, speech compressors, speech amplifiers and other complicated electronic circuits are removed, there remains the basic "CW" (continuous wave, a term used to refer to the code) transmitter—a relatively simple device. Actually, it is possible to build a CW transmitter that will work around the world—in a match box! Here's a picture of a transmitter built by Dennis Monticelli, AE6C. From his home in Fremont, California, Dennis has already used this electronic marvel to contact places as far away as South America, Japan and many other countries around the world.

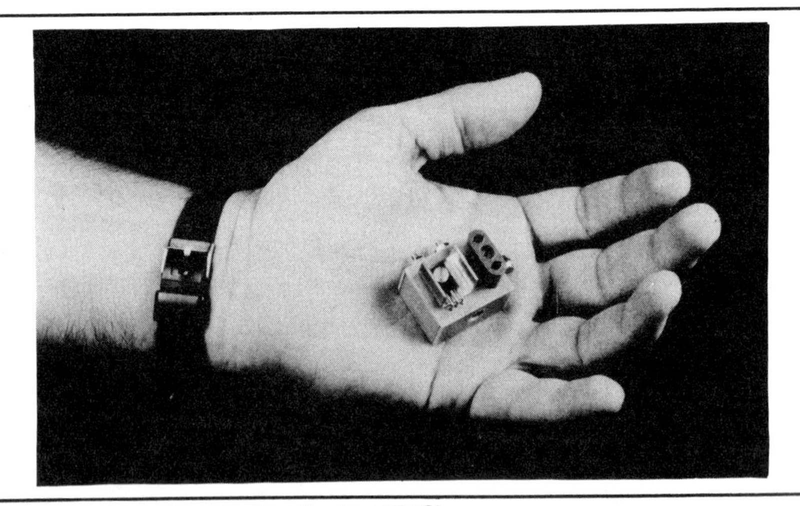

(photo courtesy Dennis Monticelli, AE6C)

The fact that a CW signal requires far less spectrum space than a voice signal is another advantage. This means many more stations can operate at the same time in any given segment of the radio spectrum. Receivers benefit from this because they need listen to only a narrow band of frequencies at any one time. By so doing, they capture less noise. And advances in electronic technology promise even narrower bandwidths for code operation. A discovery called "coherent CW" suggests that frequency spectrum usages 1/20 of that now used for CW operation are possible. This means 20 times as many code stations as before can operate in any given area of the spectrum. Or, comparing it to voice operation, several hundred CW stations can operate in the same area that is now occupied by a single FM phone station!

Furthermore, a code transmitter consumes less power than a voice transmitter. This is because current is drawn only when the key is depressed. This is especially important when battery power is used, as is frequently the case during many natural disasters, such as floods or earthquakes, when commercial power is not available. For this same reason, aircraft that are forced down at sea use battery-powered, floating beacon transmitters to send messages in Morse.

This low-duty-cycle characteristic of the code makes it especially popular for satellite communications. Radio amateurs have a number of satellites circling the earth that are used as celestial relay stations, enabling distant contacts to be made with low power. But to protect the fragile battery system of these orbiters, only those types of emission that draw minimum power are allowed. If you hope to take part in this exciting facet of the hobby some day, using the code is a good way to go.

Other types of low-current, low-power equipment also enjoy popularity in the ham fraternity. Many amateurs who enjoy camping and backpacking carry a battery-operated CW transceiver along with their other gear. Such a device is lots of fun and could prove valuable in a distress situation. No one knows this better than Jim Prior, a planning analyst from Vancouver, British Columbia.

While on a hiking trip in the Five Fingers Mountain Range on Vancouver, Jim encountered a snowstorm. He decided to descend from his 3000-foot elevation, but while he was doing so, the snow turned to rain. Conditions worsened and Jim became very cold—so cold he was barely able to function. Although he had reached a familiar lake and had access to a rubber raft, he was unable to inflate the device. As a last resort, the 39-year-old outdoorsman reached for his radio and began calling, $\overline{\text{SOS}}$... $\overline{\text{SOS}}$... $\overline{\text{SOS}}$... DE VE7CKF. He called over an hour.

Jean Ness, VE6BLJ, was in her kitchen doing chores with her radio tuned to the Trans-Canada calling frequency. As she worked, she noticed a weak, rhythmic signal buried beneath several phone signals on the frequency. She leaned closer and detected a faint plea for help. After requesting the other stations to clear the frequency, she responded to Jim's call. Upon learning of his plight, she contacted another amateur who in turn relayed the emergency message to a local air service. One and one-half hours later, Jim was safely on his way home.

In another distress-call incident, KV4FZ, of St. Croix in the Virgin Islands, spotted an $\overline{\text{SOS}}$ being sent by flashlight from a ship in distress. He notified civil defense authorities, and all passengers aboard were rescued.

And 240 miles off the east coast of the United States, a potentially tragic event developed when Ken Gaskill, third engineer of the tanker *Puerto Rican*, found a bone had lodged in his throat while eating. Fortunately, the captain of the ship, Phillips Roberts, had his amateur station aboard and put out a call for help, hoping to make contact with a medical facility that might provide assistance by radio. Although the original call was made by voice, propagation conditions deteriorated, and WA6PVB was forced to switch bands and to change modes. Using Morse code, Phillips was able to contact the Coast Guard, which instructed him to change course and head for Wilmington, North Carolina. They informed him a helicopter would be sent to meet the *Puerto Rican* in an attempt to evacuate the third engineer. Not long thereafter, the aircraft and tanker met. A wire basket was lowered from the helicopter, and Gaskill was lifted from the decks of the *Puerto Rican* and transported to the New Hanover Hospital in Wilmington. There he was given proper medical treatment and eventually released.

But emergency purposes aside, the code has still more advantages. It can open many doors for you in both your vocational and avocational pursuits. By becoming proficient in its use, you will master one of the requirements for obtaining a commercial radiotelegraph license or Amateur Radio license. You will understand and enjoy the thousands of CW signals that cover the shortwave spectrum.

Also, if you are interested in computers, be advised that the code has kept pace with technology. Electronic keyers, CW keyboards and home computers are routinely used to transmit Morse. Reception is often accomplished by displaying messages on LED readouts, communications terminal screens, home-computer CRTs and dot-matrix printers. A concept called "Super-CW" makes use of computers at both ends of a transmission path and promises to increase message accuracy considerably.

And if you enjoy competing in contests, you'll never be bored with the code, for there are many amateur activities for the CW enthusiast. To name just a few, there are the ARRL DX Contest, ARRL November Sweepstakes, All Asian CW Contest, various 160-meter CW contests, CQ World-wide DX Contest, Straight-Key Night and Code Proficiency Runs.

But if all the reasons given in this chapter to use the Morse code didn't exist, there would still be one reason for its use that is strong enough to stand by itself... The code is just plain fun!

CHAPTER TWO

A History of Telegraphy

The transmission of messages by prearranged signals dates to the earliest of times. Signal fires were used by the Romans; Greek Army commanders used polished shields that reflected the sun to transmit messages of war; the Indians communicated with columns of smoke; and drums have been used by primitive nations for years. Some say the initial concept of the Morse code can be traced to the third century BC, when the Greek historian Polybius used an array of torches to transmit signals during the Punic Wars. These torches of light were alternately hidden and displayed to convey their messages.

1792...

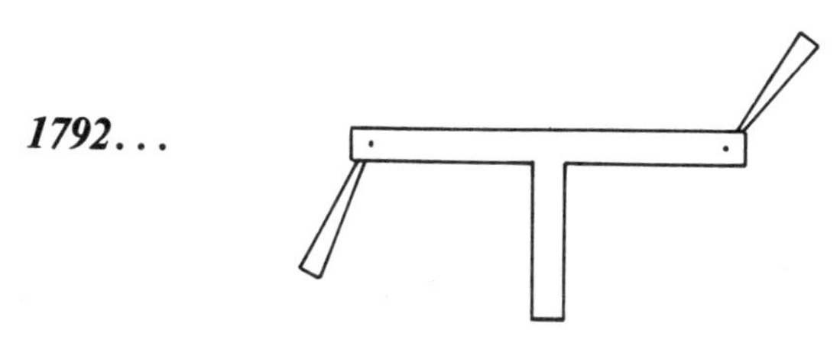

The invention of the first true "telegraph" didn't take place until 1792, however, when French inventor Claude Chappe devised a mechanical contrivance that used two movable arms attached to a post. By means of an early form of semaphore codes, this machine could transmit messages to an observer miles away using a telescope to view the position of the arms. Each arm assumed one of seven different positions as it was rotated around the main supporting beam. This gave a total of 49 combinations of arm positions, which were used to indicate various letters and numerals.

This invention proved to be far superior to the conventional courier method used at the time. With stations set up at intervals of three to six miles apart, 100 times as many messages could be transmitted from Toulon to Paris, for example, as were previously carried by courier. The device was used extensively in the 1790s to transmit messages all over France, and it was during this period that the word *telegraph* was coined.

1819...

Improvements were made to various types of mechanical units, but electric telegraphic concepts did not begin to emerge until around 1819. In that year, Hans Christian Orsted, a professor of physics at the University of Copenhagen, learned that a magnetized needle could be deflected when an electric current was passed through a nearby wire. This fact led to the development of several "needle-telegraphs" operating on this newly discovered principle. In 1832, a needle-telegraph was in operation at St Petersburg, Russia, and a similar German device was in use as early as 1833.

The first patented electric telegraph was built in 1837. A partnership was formed in London between Charles Wheatstone, a British physicist, and William Cooke, an ex-Indian Army officer, for the purpose of introducing and developing an electric telegraph for the English railway system. Their device was a six-wire telegraph that used five magnetized needles. By application of an electric current to different combinations of wires, two needles could be made to point to a particular letter or number that was drawn on the face of the telegraph.

Unfortunately, the railway officials to whom the machine was demonstrated were not impressed. Cooke and Wheatstone had difficulty proving the worth of their idea. But then, in 1844, a strange event occurred that brought this new contrivance to the public eye. A woman was murdered in Salthill, England, by a man who boarded a train for London shortly thereafter. The police in the town of Slough, near Salthill, telegraphed a message to the London police, telling of the event and describing the man. Upon his arrival in London, the perpetrator, John Tawell, was apprehended. He was later tried, convicted and hanged. The news of the event made headlines and confirmed the usefulness of this new device.

1831...

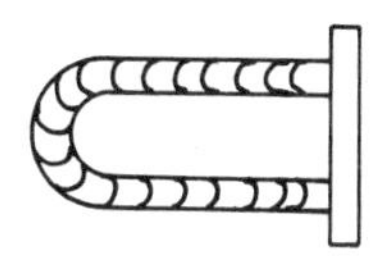

During the time that Cooke and Wheatstone were developing their needle-telegraph, a teacher at Albany Academy in New York, Joseph Henry, was experimenting with electric magnets of two different types. One used an iron core wound with a few turns of wire through which a heavy current was passed. Another used a core wound with many turns of wire carrying a smaller current. Henry eventually constructed the most powerful magnet ever devised until that time. It was capable of supporting a weight of almost two tons.

Henry was also interested in telegraphy and how he could apply the principles of electromagnetism to its use. In 1831, he devised an electromagnetic telegraph instrument that used a magnetized bar mounted on a centralized pivot. One end of the bar was placed between the ends of an electromagnet and the other end was positioned near a bell. When current was applied to the electromagnet, it would move the bar, which in turn hit the bell. Henry was able to energize this device through a mile-long length of wire, and he used one of these instruments to communicate from his house to his laboratory at Princeton University. This was truly a great step forward.

1832...

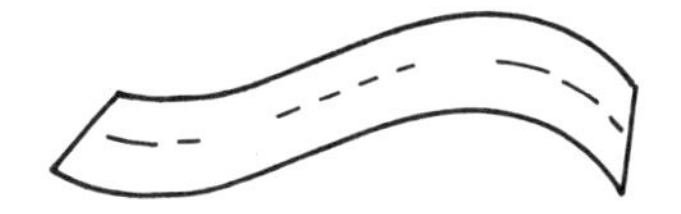

The name Samuel F. B. Morse is synonymous with telegraphy, and so it should be. However, few realize that Morse's genius extended beyond the commercial development of the telegraph and the invention of the code, and that it was actually in the field of *art* that he first became well-known.

Morse was born in Massachusetts in 1791, the son of a minister. He graduated from Yale College in 1810 and left for London in 1811 to study art. After several years as a struggling artist, his work began to be recognized. In 1813 and again in 1815, some of his paintings were accepted into the Royal Academy art exhibit. Others of his works were later to be exhibited in the New York City Hall and New York Public Library. He became a great American artist.

Morse's interest in telegraphy began in 1832 while homeward

bound from Europe. On board the ship *Sully*, he became engaged in a dinner conversation with some men who were discussing how it was possible to send currents of electricity through any length of wire instantaneously. Morse immediately became interested, and it was at that point that the telegraph became his all-consuming preoccupation. He spent the remainder of his voyage making notes and drawing diagrams.

Upon arriving home from Europe, Morse found himself almost penniless. It was necessary for him to live with his brothers while he continued work on his invention. He earned some money by teaching sculpturing and painting, but every dollar went toward the purchase of supplies for a telegraph model.

Morse worked hard. It was impossible in those days to buy long lengths of insulated wire, so he had to manufacture his own. Painstakingly, he soldered short lengths of bare wire together and wrapped them by hand with cotton thread. Working night and day, he fabricated miles of wire needed for the project.

His transmitting device consisted of a long, thin, wooden tray called a port-rule, which contained notched metal pieces. This was drawn under a set of electrical contacts, and as it moved, would cause the contacts to open and close, thereby alternately opening and completing an electrical circuit. At the opposite end of 1700 feet of wire was the receiving unit. It recorded the dot-dash type message that was notched in the port-rule by writing marks with a pen onto a strip of moving paper. The movement of the pen was controlled by an electromagnet that responded to the electrical impulses passed through the wire, while the paper tape was drawn by a clock mechanism. Morse used a codebook at the transmitting and receiving end. Various codes were used to indicate certain words, names, dates, and so on.

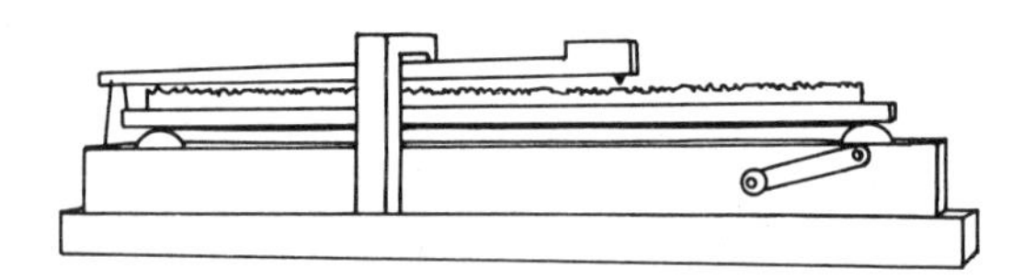

Although this device operated successfully, it had drawbacks. The use of a codebook was cumbersome and severely limited the contents of the transmitted message. Also, the notched port-rule was awkward and slow. A faster, more flexible means of encoding and decoding was needed. Years later these problems led to the invention of the Morse code and to the use of a simple open-close switch (key).

Morse demonstrated his port rule mechanism in 1837 to a group of influential men, but little interest was shown. Only Alfred Vail from Morristown, New Jersey, expressed interest. Vail agreed to become a one-quarter partner in Morse's invention and helped him build a sturdier model. Even with his new partner, however, he was unable to market the device successfully. Years of disappointment followed.

In 1842, an impressive demonstration was arranged in which several miles of wire were laid underwater from the Battery to Governors Island in New York. Just before the demonstration took place, a ship caught the telegraph wire in its anchor. When the anchor was raised, the ship's crew cut the wire, not realizing what it was. Accounts of Morse's failure to contact the island suggested a hoax.

Finally, Congress appropriated money for a major test of Morse's instrument, and in 1844, a 37-mile stretch of line was constructed from Washington to Baltimore. On May 24 of that year, Morse sat in the Capitol Building in Washington and sent his famous message—"What hath God wrought." The receiving unit in Baltimore recorded the transmission on a strip of paper tape.

Working independently, others originated a variety of sending and receiving devices in an attempt to improve upon the use of Morse's code. Several automatic chemical telegraph systems resulted from some of these efforts. At the transmitting end, such systems made use of paper tape that was pierced with holes and slits that represented dots and dashes in the Morse alphabet. This was fed into a tape-reading device that moved it at constant speed past a spring-loaded contact mechanism that alternately made and broke electrical contact as it encountered perforations in the tape. The receiving side of the unit employed another tape that had been immersed in a chemical solution prior to use. This tape passed between a rotating, metallic cylinder and a metal pen or stylus. Electrical impulses from the transmission line were applied to the pen and cylinder. Current that passed from the pen to the cylinder decomposed the chemical solution in which the paper tape had been immersed. This action caused a discoloration of the tape, producing marks that coincided with the transmitted Morse characters.

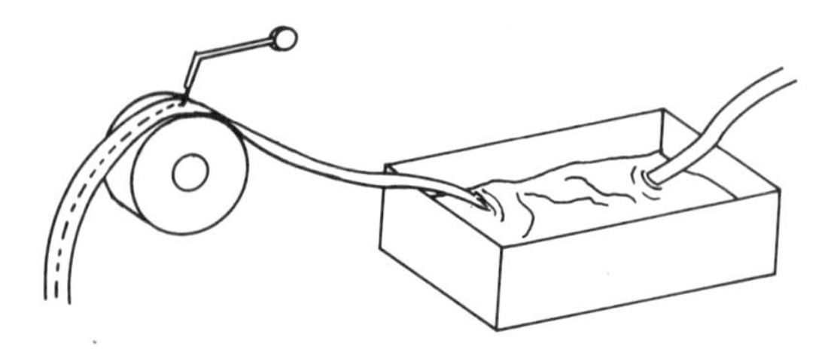

Charles Wheatstone developed a system that made use of a similar type of transmitting mechanism, but employed an ink-recording device at the receiving end. The transmitting tape, commonly called a "slip," was prepared on a perforating machine and fed through a tape-reading unit similar to that used by the chemical telegraphs. At the receiver, another tape was driven past a magnetically operated, inked roller that responded to the transmitted impulses and jotted the coded message onto a receiving slip.

Using the automatic chemical and Wheatstone systems, messages could be transmitted at much higher rates of speed than by hand. Because of this, a number of tape perforators were required to feed a single system. Although arrangements allowed for theoretical speeds up to 600 words per minute, electrical characteristics of transmission lines limited actual operation to rates far below that figure.

Use of the Morse telegraph spread rapidly. Soon, lines dotted the United States and private enterprises utilizing the invention began to appear. By 1851, there were more than 50 telegraph companies. In 1856, the Western Union Telegraph Company was formed from the combination of 12 other companies. Beginning with 132 offices in that year, Western Union operations expanded so fast that it was operating 4000 offices by the year 1866.

It is interesting to note that both Morse's original telegraph and the one used in his famous May 24, 1844 demonstration recorded their messages on paper tape. Later models, however, used a "sounder." The sounder was an electromagnet that received impulses transmitted by a telegraph key at the opposite end of the line. As dots and dashes were sent, the sounder would close and open, making a clicking sound. Exactly when the conversion was made from recording tape to the sounder is not clear. One account indicates that an unknown operator was monitoring the visual output from a message on tape around 1850. Then he unconsciously began copying the coded letters in his head while listening to the sound of the pen, and began writing each letter down by hand. Suddenly, this new, faster method of reception became apparent. It eventually became the most widely used means of receiving code at the time.

This was the heyday of the "landline" telegraph, but there were still many problems to overcome. Telegraph lines were expensive to manufacture, install and maintain. The single-circuit method of transmitting code was slow. An average operator could transmit at speeds of 20 to 25 words per minute and an exceptional one at 30. Eventually this problem was overcome by the invention of multiplexing.

1861... War!

No discussion of telegraphy would be complete without mention of the Civil War, for it played an enormous role in that conflict. Operators enlisted in large numbers, and the telegraph key and sounder became standard tools of war.

Starting with four top-notch operators formerly with the Pennsylvania Railroad, the North eventually enlisted 1200. However, their numbers dwindled to only 200 by war's end, for it was dangerous duty. Operators had to advance to the front line and stay put until the last minute during times of retreat so news could be relayed to the rear guard. Many were only 12 years old.

They would lay line on the ground for temporary installations and attach it to poles for more permanent use. Typically, a wagon carrying all necessary supplies of poles, wire, etc, would cover 20 miles in 8 hours. A team of men installed 12-foot-long poles into the ground at intervals of about 130 feet. All told, some 6,000,000 messages were transmitted over 15,000 miles of wire during the hostilities.

1872...

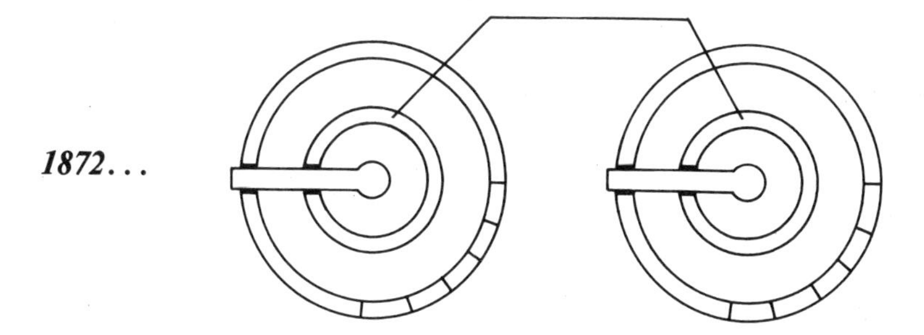

Emile Baudot, a French telegrapher, provided a giant step forward in 1872 with the invention of multiplexing, a system that allowed numerous messages to be sent over a single line simultaneously.

The basics behind multiplexing provided for a rotary mechanism at the transmitting end synchronized with a similar mechanism at the receiving end. At a particular point in the rotation of these units, a telegraph line would be at the disposal of one pair of operators. At another point of rotation, a different pair of operators had use of the same line, and so forth. To accomplish this feat, a stationary faceplate fitted with concentric copper rings was used. Then, brushes mounted on an arm that revolved around the center of the faceplate made contact with signal points from different operators as they rotated. When officially adopted by the French Administration of Posts and Telegraphs in 1877, the Morse/Baudot converter system permitted use of a single line by four sets of telegraphers. Later enhancements to the system allowed transmission of many more signals.

1831...

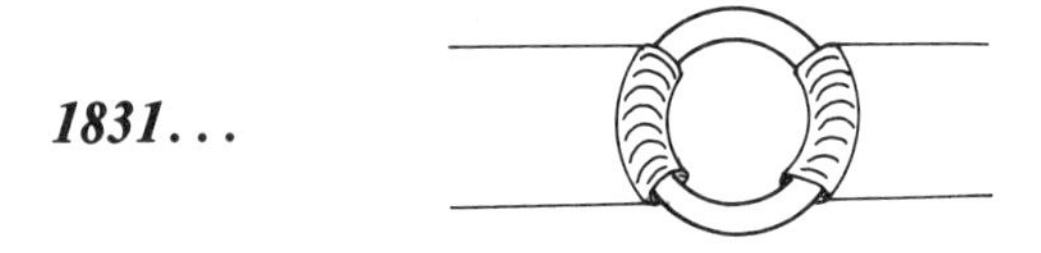

Advances continued to be made in the landline telegraph system. At the same time, exciting discoveries were emerging from experiments with electricity that would later lead to use of the word "wireless." (Notice we have backed up to 1831 to cover these innovations from the beginning.)

It was already known that electric currents could produce magnetism, and this knowledge was used in the early needle-telegraphs. But many wondered if the reverse was possible. Michael Faraday, an English physicist and one of the most brilliant experimenters the world has ever known, addressed the question.

In his attempts to create electric current from magnetism, Faraday tried many things with little success. Finally, in 1831, he accomplished what he had years earlier set out to do. His triumph was first demonstrated by using a circular iron ring wound with two separate coils of wire on opposite sides. One coil was connected to a battery, the other across a galvanometer (an instrument used to measure electric current). Nothing happened until the battery circuit was alternately switched open and closed, at which time the galvanometer was deflected by an induced current. Another demonstration was set up similarly, but used an iron cylinder with coils wound at opposite ends. A third used a single helical coil of wire through which a magnet was passed. All showed galvanometer deflections. His later experiments proved current could be induced without the aid of an iron core. These monumental discoveries laid the groundwork for the wireless age that was to come.

1842...

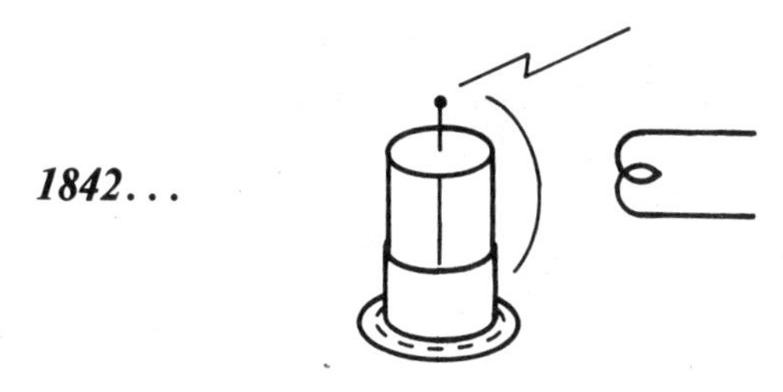

Joseph Henry, who developed the electromagnet in 1831, was involved in further experiments with magnetic induction. In 1842, he discovered that the discharge of a Leyden jar was oscillatory. That is, it discharged first in one direction, then the next, alternating again back and forth, each discharge being weaker than the one before—the first indication of discoveries that would lead to use of the word "frequency."

Henry made additional exciting findings. During one investigation, he learned that a 1-inch spark from a Leyden jar discharge at the end of a length of wire in his room induced a current in a parallel length of wire 30 feet away in his basement.

1865...

Stories of the first true wireless transmission through space vary. It is generally accepted that this feat occurred in 1865, during an experiment that is reminiscent of the famous Benjamin Franklin kite-flying story. In the summer of that year, Mahlon Loomis, a dentist from Washington, DC, set aloft a kite from a mountaintop in West Virginia. The surface of the kite was covered with a square of copper gauze. Its string consisted of a length of fine copper wire connected through a galvanometer to a coil of wire buried beneath the earth. A similar kite with the identical arrangement was flown by a group of his associates on another mountain top. Then Dr. Loomis broke the copper wire connection to his galvanometer; at the same instant, 18 miles away, the galvanometer connected to the other kite moved!

A & P Telegraph Company key—1850

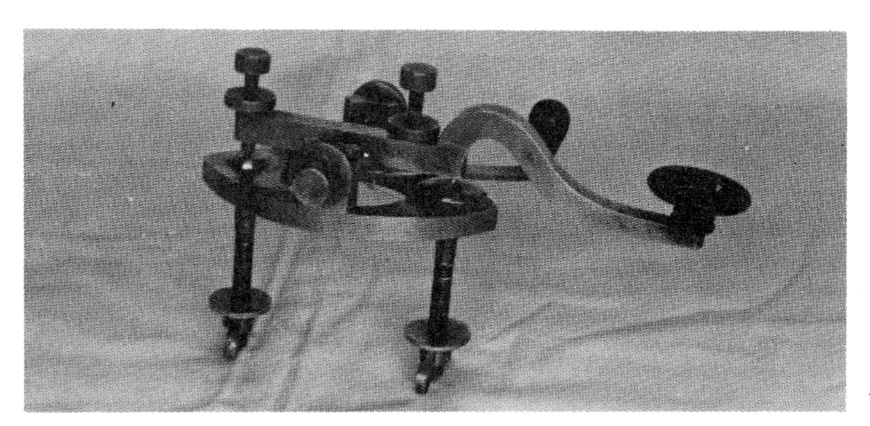

"Camelback" key used during Civil War—1860

Western Electric Company key—1875

All keys are from the collection of Louise Ramsey Moreau, W3WRE.

Mahlon Loomis never received recognition for this feat that was in line with its importance. He does go down in history, however, as the man who first used, and named, the "aerial."

1886...

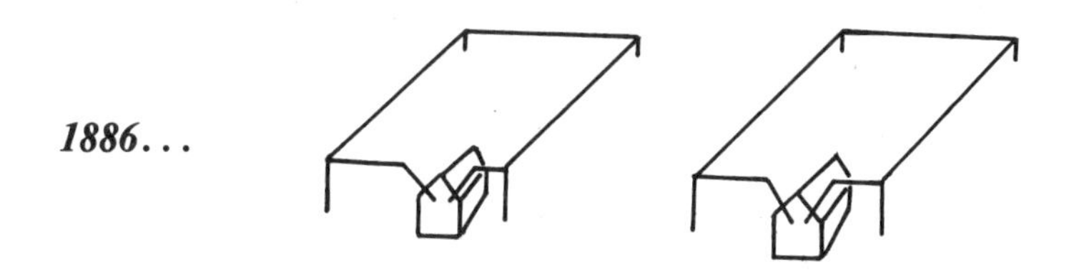

One of the earliest attempts at wireless communication over an appreciable distance was made by William H. Preece, a British electrical engineer. In 1886, Preece set up two stations one-quarter mile apart in Newcastle, England. Each station used a square antenna that was 440 yards long on one side. Contact between the two was accomplished easily. After this successful attempt, the stations were moved farther apart. Again contact was made at a separation of more than one-half mile. It was finally determined, however, that the signal strength from the transmitting unit diminished appreciably when stations were separated by a distance that was greater than the total length of a single antenna.

1887...

One year after Preece's experiments, a German physicist, Heinrich Hertz, demonstrated the existence of electromagnetic waves. Hertz used two spark gaps (a pair of conductors separated by a short distance) about 5 feet apart. Each spark gap was centered at the focal point of a metal parabolic mirror. When a spark was made to jump across one gap it caused a spark to jump across the other also. By aiming the mirrors in various directions, he showed that electromagnetic waves traveled in straight lines and could be deflected by a metal surface. Additional experiments demonstrated the properties of these new waves to be in accordance with those of light and heat waves.

Hertz performed other experiments with cathode rays in a vacuum that bordered on the world of the atom and electron. When he died at the age of 37, this brilliant physicist left behind epochal work that would be continued by later pioneers in the wireless age.

1892... 

Two years before Hertz's death, a French physicist named Edouard Branly completed a device that he called a "coherer," considered the final link in the chain that led to successful wireless communication. This invention consisted of a glass tube with two silver plugs inside that were connected to an external receiving circuit. A small amount of metal filings was in between the plugs. Normally, this device exhibited a very high resistance to electric current; however, when an electromagnetic wave was introduced nearby, the resistance of the coherer decreased markedly. Because of the way it reacted, this unit served as a wave detector. It had drawbacks, one being that the tube had to be tapped each time after it detected a wave for the metal filings to fall back into their original position and make the coherer usable for the next wave burst that came along. Despite its shortcomings, the device worked.

Many names from many countries have been mentioned in this long list of discoveries—Faraday, Henry, Morse, Loomis, Preece, Hertz—England, America, Germany and France. With all these inventions, hard-earned over so many years, someone was needed to tie them together and make them work for the ultimate purpose. Who would that be?

1896... At Last

Guglielmo Marconi, son of an Italian father and Irish mother, was born in Italy on April 25, 1874. He was fascinated by electricity and intrigued by the experiments of his predecessors. The story is told that he first became intensely interested in wireless at the age of 18, after reading an essay in the *Fortnightly Review* by Sir William Crookes, a British physicist. Crookes suggested the possibility of wireless, indicating all that was needed was a combination of some means of generating waves of any desired wavelength, an improved coherer and a method of directing transmission and reception. If each of these items was properly designed and made to work in conjunction with one another, "telegraph across space" would be possible.

Ideas raced through his head as he thought of ways to accomplish such a feat.

Working with crude apparatus, Marconi used a coherer as the receiving unit and a Morse telegraph key in conjunction with a spark discharger and induction coil for transmitting. He experienced enough success over short distances to experiment further. After improving his coherer, he found that attaching a vertical antenna to the equipment made transmission over greater distances possible. He even tried a combination of wires in an attempt to create a deflector for his vertical antenna that he hoped would improve communication. Eventually, signaling range was increased to one and one-half miles. The year was 1895.

After attempting without success to interest Italian authorities in his findings, Marconi went to England, where he received, in 1896, a patent for his invention. In addition, he met with and was assisted by Sir William Preece who helped promote the new wireless idea. A variety of demonstrations were set up at different points around England, some of which included the use of kites and balloons to raise aerials high into the air. By so doing, transmission between points up to nine miles apart was achieved.

After these successes, Marconi was received with open arms by various British interests. Only 20 years before, they had capitalized on the introduction of the telephone and now they pictured even greater profits. In 1897, the Wireless Telegraph and Signal Company, Ltd, was formed. Three years later, its name was changed to Marconi's Wireless Telegraph Company, Ltd. Not forgetting the cold reception his idea received in his native land, Marconi assigned patent rights worldwide—for every country except Italy.

The principal purpose of this new company was to promulgate the use of Marconi's wireless. Continuous attempts at new and better installations resulted in transmission from South Foreland, England, to Wimereux, France, in 1899, a distance of 31 miles. In the same year, three British warships were equipped with apparatus that enabled them to transmit over distances of 75 miles. On April 28, 1899, the new wireless, still in its infancy, demonstrated its usefulness during emergencies by saving the lightship *Goodwin Sands* off the English coast. A severe storm badly battered the ship, which reported its condition to a nearby telegraph station. Another vessel was dispatched to its aid, and loss of life was prevented.

Toward the end of 1899, Marconi went to the United States, where he carried out more amazing demonstrations. In September, two US ships were outfitted with wireless and used the gear to relay news of events in the America's Cup yacht race to newspapers on shore. The world applauded. Shortly thereafter, the American Marconi Company opened its doors for business.

December 12, 1901 marked an event as significant as the first wireless transmission itself. Marconi had returned to England and built a powerful transmitting station at Poldhu in Cornwall. When it was finished, he traveled to St John's, Newfoundland and erected a receiving station on Signal Hill that contained the latest in wireless equipment including an antenna that was to be supported by a kite at an elevation of hundreds of feet. All this was in preparation for an attempt to span the Atlantic with electromagnetic waves.

At Signal Hill, two attempts to raise the antenna failed. In the first, a 9-foot, bamboo and silk kite was tried, but the wire snapped and the kite drifted out to sea. Then a balloon was tried, and again the line broke. Finally, a second kite was used to successfully raise the receiving antenna to a height of 400 feet. And then he listened. Through the static came the prearranged signal from Poldhu, England—a raspy "S" sent in Morse code. It was heard again, twice, on that afternoon of December 12.

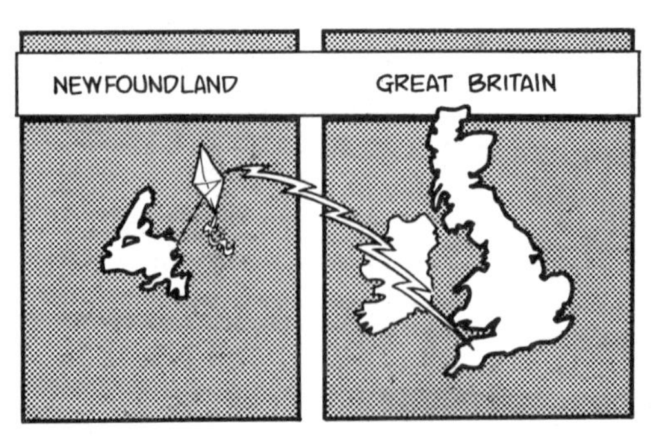

When the news was released, the world was astounded! The initial reaction was one of disbelief. How could such an impossible event have occurred? But it was true, and it was proved again the following month when the liner *Philadelphia* picked up Poldhu's signal from 2099 miles away. Marconi was 27 years old at the time.

Enter the Amateur

The true Amateur Radio operator, in the conventional sense, began to appear around 1900. Perhaps Amateur Radio *experimenter* would be a better phrase to describe this turn-of-the-century radio enthusiast, for there was little real radio communication between station operators at the time. But Marconi's success fired the imaginations of many young people—and the movement began.

At this point, it was difficult to separate the words amateur and professional because many of those who made contributions to the newborn field of wireless began as amateurs but eventually became professionals. Also, because the science was so young, an amateur who made significant contributions might become a professional overnight. The separation between these terms grew more pronounced as time went on.

The age of wireless had begun, and everyone was talking about it. Stories appeared daily in newspapers from coast to coast. The value of on-the-air communication proved itself over and over again both for commercial use and during times of sea disaster. But there was still a great deal of work to be done.

Communication was unreliable, subject to the vagaries of atmospheric conditions, weather and the time of day or night. Distortion and fading were the rule rather than the exception. Spark-gap transmitters emitted a very "broad" signal. That is, a wide range of the frequency spectrum was taken up by an individual transmission, making it impossible for two stations operating near each other to be on the air simultaneously.

These problems were addressed by several pioneers in the field. One of the initial steps in overcoming these difficulties, and indeed one of the first steps leading the way into modern day electronics, was taken by Sir John Ambrose Fleming in 1904. Fleming, one of Marconi's assistants, was experimenting with the "Edison effect" in an attempt to find a way to convert the oscillating current received by

American DeForest spark key—1909

"The Boston Key"—1915

Signal Electric Company spark key—1918

All keys shown are from the collection of Louise Ramsey Moreau, W3WRE a transmitted signal into a steady current that could deflect and hold a galvanometer needle.

The Edison effect was a phenomenon noticed by Thomas Edison when he was working with his incandescent electric lamp. A battery attached across the filament of a lamp made the filament glow, producing light. But after a period of use, the inside of many bulbs became darkened with what appeared to be a "soot," which cut down on the light output. Some bulbs, however, did not have this layer of soot in a thin area adjacent to the edges of the filament. Attempting to learn more, Edison inserted a small metal plate inside one bulb, near the filament but not touching it. He was shocked to find that with a galvanometer inserted in line between the positive side of the battery and this plate, current flowed between the plate and filament.

Fleming, using Edison's findings, experimented further. The result was the invention of the first "diode," a device that allows current to flow in one direction but not the other. Thus, Fleming's research made it possible to "rectify" an oscillating (alternating) current and change it into a steady (direct) current. Electromagnetic waves that passed from transmitting antennas to receiving antennas could be rectified in the same manner so a steady indication of the transmitted signal could be obtained. In other words, the signal could be "detected."

This momentous discovery initiated a long series of future developments in electronics and telegraphy—too numerous to cover in detail. But there is one other that followed directly on the heels of Fleming's invention that is equally, if not more, important. This epochal discovery was made by the "father of radio," Dr. Lee DeForest.

DeForest attended the Sheffield Scientific School at Yale and upon graduation continued his education, receiving his PhD from Yale in 1899. He began work in wireless telegraphy at the Western Electric Company, and his efforts expended on hundreds of experiments resulted in the invention of the "audion." This was a vacuum tube, similar to the Fleming diode, with one exception. It contained an additional element called a "grid." The grid was placed between the filament and plate in Fleming's diode, and was connected to an antenna. This new piece of electronic equipment proved to be not only a better detector of radio waves, but also an amplifier.

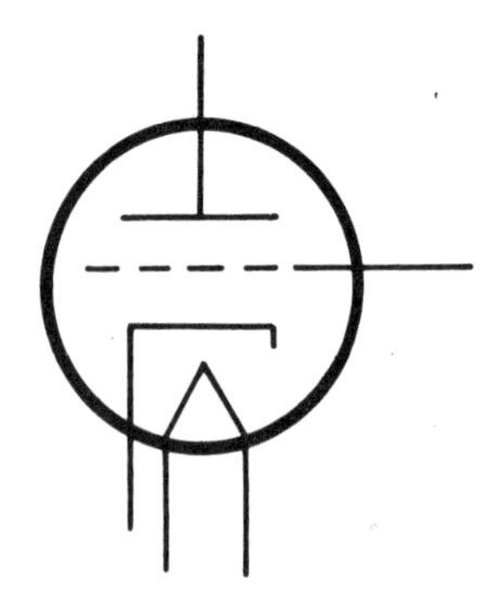

Although even Lee DeForest himself did not appreciate the potential of his idea at the time, the audion eventually proved itself to be of immense importance in radio communications. It was, in fact, the original vacuum tube, and it later became the backbone of *every* wireless set made.

By this time, there were perhaps several hundred amateur stations on the air. "Home-brew" was the fundamental principle of the day. Almost nothing was store-bought; equipment was difficult to come by and expensive when found. The amateur station was truly homebuilt.

Primary emphasis was placed on improving reception. DeForest's audion was used initially, but its expense prevented most amateurs from owning one. A device called an electrolytic detector became popular because it could be homebrewed. It remained in use for years until replaced by the crystal detector.

Some advances were made in transmitting equipment, although the spark gap remained the backbone of the amateur's station. Improved spark gaps were developed that used a shorter and wider spark, a quenched gap, an enclosed gap and eventually a rotary gap which, according to advertising literature of the day, "increased efficiency 25% and produced a very high, clear note." To a contemporary listener, that beautiful note of yesteryear might have closer resemblance to a jamming signal than that of a bona fide CW transmission. But it was progress, nevertheless.

In 1909, the first Amateur Radio organization was formed in New York City, the Junior Wireless Club, Limited. The book *Wireless Telegraph Construction for Amateurs* was published the following year, and amateur organizations began cropping up around the country. In May 1914, the American Radio Relay League was formed. If there is any doubt as to the importance of Amateur Radio in the development of radio communications, take note that the amateur of the early 1900s had equipment superior to that of the United States Navy! For more information, see *200 Meters and Down*, published by ARRL.

Of major importance during this period was the formulation of a communications law. Prior to the signing of the Alexander Bill on August 17, 1912, the ham operator had as much right to the airwaves as anyone else. Battles between amateurs and commercial stations were common—and the amateurs usually won! With the passage of this bill, the amateur was "put out of commission" by being restricted to the "useless" portion of the radio spectrum on wavelengths shorter than 200 meters (higher in frequency than approximately 1.5 MHz) and above. In those days, it was believed that anything above this frequency was worthless for communications purposes.

Such a restriction, although it seemed severe at the time, could not hold back the enthusiasm of this group that now numbered nearly 4000. Experiments continued, and with improvements in the vacuum tube, progress was made. Edwin H. Armstrong's name stands in the forefront as the amateur who vastly improved the detection and amplification characteristics of the tube, resulting in a giant leap forward for radio communications.

Another event made dramatic Amateur Radio history during World War I, when Charles E. Apgar, 2MN, from Westfield, New Jersey, designed a sensitive receiving circuit and began experimenting with the recording of radio transmissions. A commercial station, WSL, located on Long Island in New York, came under suspicion as possibly being involved with clandestine enemy war activities. The chief of the US Secret Service asked L. Krumm, Radio Inspector at the Customhouse in New York, to investigate the matter. Krumm had recently learned of Apgar's recording experiments and asked the amateur to record WSL's late-night transmissions. This was done, and daily the recordings were rushed to the Secret Service office for study. Analysis showed that the station was transmitting sensitive information to the Germans concerning Allied shipments on the high seas, and station WSL was immediately closed down. Charles Apgar had performed a truly monumental service to his country.

Amateurs continued to build, experiment, discover and invent. Perhaps the most significant contribution to telegraphy since Marconi's historic transmission came about with the inception of vacuum-tube CW transmitters. The emission from these "newfangled" devices occupied an area of the spectrum that was less than 1 percent of that required by a spark transmitter. As welcome as such an innovation may appear, the new means of sending code did not gain overnight acceptance because there were problems involved. Receivers of the day could whisk by such a sharp signal and never detect it, and transmitting tubes were hard to find and ran low power. But, as with other obstacles encountered, the amateur eventually overcame these problems, and tube transmitters proved to be far superior to their earlier counterpart.

Voice communication was in use by this time, but the code continued to be the only reliable means of two-way contact over extended distances. Although the 1920s saw the beginning of an era of commercial broadcast 'phone stations, the code maintained a major role in communications efforts up to and throughout World War II. During this period, amateur communications by Morse code made history not only because of technological advances but also because of the amateur's role in civilian public service activities and the war effort.

A review of the record shows that in November 1930, amateurs proved to be the only reliable means of communications when a severe sleet storm hit the Midwest. The fraternity responded by handling emergency traffic on code for stranded Westerners, the Union Pacific Railway and the Postal Telegraph Service.

Also, hams were there in March 1936 when one of the worst floods in history struck the eastern United States. Heavy rains and thawing ice swelled rivers and streams. Entire towns were inundated. Hundreds of amateurs responded, coordinating efforts on both CW and phone channels, lending assistance in rescue operations for more than a half million Americans left homeless.

And in December 1941, when war broke out, amateurs enlisted by the thousands in the Army, Navy and Marines, using their expertise in telegraphy to aid the war effort. Some studied Japanese and Russian telegraph codes. Others participated in civil defense communications. Still others studied cryptography, applying their skills to creating and breaking secret codes.

After the war, rapid advances were made in code sending and receiving devices. Transistorized equipment, integrated circuits and other sophisticated electronics found their way into CW gear. The straight key and sounder were replaced by high-speed telegraph machines, electronic keyers, CW audio filters, automatic code readers, CW keyboards, computer interfaces and satellite relays.

In addition to modern-day use by radio amateurs, the code continued to be utilized by commercial enterprises and the military. Its primary usage is found today in maritime communications; ship-to-ship and ship-to-shore messages are regularly transmitted in Morse.

Listening to some of the maritime bands listed in Chapter 4, one might hear weather reports in Morse, or orders for food from a luxury liner, lists of passengers, reports of parts needed to repair on-board equipment failures, requests for medical supplies, pleas for assistance from the Coast Guard, warnings of iceberg sightings or simply idle chatter from off-duty operators.

Communications on Navy frequencies could include accounts of war games, reports of foreign military vessel sightings, messages to submarines, encrypted traffic or live military operations from hot spots in various parts of the world.

The code is also very much in use commercially and militarily on frequencies below the AM radio band. Here, in addition to maritime communications, thousands of CW beacons dot the airways, operating 24 hours per day. These electronic lighthouses are used on land and sea to pinpoint the locations of airports, offshore drilling platforms and other points of reference for air and sea traffic.

The means of sending and receiving Morse code have changed, but its usefulness remains the same. It continues to be the most reliable form of communications ever devised. Recent stories of its use under a variety of circumstances were related in Chapter 1. Other stories appear in newspaper and magazine articles. For example:

February 1976...WA1PSI notified the Coast Guard after he intercepted an $\overline{\text{SOS}}$ from a ship on fire in the Atlantic Ocean.

February 1976...WA6INJ, unable to speak because of the extreme cold, saved his life by tapping out Morse code with his microphone push-to-talk button after his jeep went over a cliff during a snowstorm.

February 1976...The ketch *Spellbound* sent an $\overline{\text{SOS}}$ that was received by KH6DE and KH6EW, who notified the Coast Guard.

March 1976...WA1PSI received an $\overline{\text{SOS}}$ from a ship in the

Atlantic that had lost its navigation gear.

May 1980...The Coast Guard was notified when KA5FYI, KH6JTI and WD6FKT heard an $\overline{\text{SOS}}$ from a ship on fire and sinking near Costa Rica.

August 1983...AG3L used his battery-powered CW transmitter to maintain communications during his solo sea journey from Chesapeake Bay to Bermuda and back.

November 1988...One hundred miles off the coast of Florida the *U.S.N.S. Bartlett* went dead in the water. Since the main power system was down, its high-technology satellite terminal became useless. Radio Officer Richard J. Monjure (N5JOB) put out a call in Morse code from a 1940's vintage, 40-watt emergency transmitter operated by battery power. The day was saved when a prompt response was received from another vessel located 350 miles from the *Bartlett.*

August 1989...The sailing vessel *Blue Agean* put out a CW distress call when it was caught in gale force winds east of Halifax, Nova Scotia. The Coast Guard and several ships responded.

And the stories will continue...You may be part of the next one. Will you be ready?

CHAPTER THREE

The Code

There are two telegraph codes, and sometimes their names are confused with one another. This book is devoted to the International Morse Code, sometimes called the International Code, Continental Code, Radiotelegraph Code or just plain Morse code. It is used almost exclusively throughout the world today. This code, however, was derived from an earlier version of the Morse code known as the American Morse Code. Although use of American Morse is rare today, it is interesting to compare this code with the newer version.

Both codes are comprised of letters, numbers and punctuation marks that are represented by a combination of dots and dashes that will hereafter be referred to as "dits" and "dahs." This is because the correct way to learn the code is by the sound of each character, not by memorizing the fact that a particular character is comprised of a certain series of dots and dashes. For instance, from the standpoint of how the letter N sounds, it is easier to recognize it as "dah-dit" than "dash-dot." This will be covered in greater detail in the next chapter.

In the code that we shall study in depth, International Morse, the duration of a dah is three times as long as that of a dit, and the space between dits and dahs in any particular character is equal to the duration of one dit. The space between characters is equal to three dits, and the space between words is equal to seven dits.

Character	INTERNATIONAL MORSE	AMERICAN MORSE
A	• ▬	• ▬
B	▬ • • •	▬ • • •
C	▬ • ▬ •	• • •
D	▬ • •	▬ • •
E	•	•
F	• • ▬ •	• ▬ •
G	▬ ▬ •	▬ ▬ •
H	• • • •	• • • •
I	• •	• •
J	• ▬ ▬ ▬	▬ • ▬ •
K	▬ • ▬	▬ • ▬
L	• ▬ • •	▬▬▬
M	▬ ▬	▬ ▬
N	▬ •	▬ •
O	▬ ▬ ▬	• •
P	• ▬ ▬ •	• • • • •
Q	▬ ▬ • ▬	• • ▬ •
R	• ▬ •	• • •
S	• • •	• • •
T	▬	▬
U	• • ▬	• • ▬
V	• • • ▬	• • • ▬
W	• ▬ ▬	• ▬ ▬
X	▬ • • ▬	• ▬ • •
Y	▬ • ▬ ▬	• • • •
Z	▬ ▬ • •	• • • •
1	• ▬ ▬ ▬ ▬	• ▬ ▬ •
2	• • ▬ ▬ ▬	• • ▬ • •
3	• • • ▬ ▬	• • • ▬ •
4	• • • • ▬	• • • • ▬
5	• • • • •	▬ ▬ ▬
6	▬ • • • •	• • • • • •
7	▬ ▬ • • •	▬ ▬ • •
8	▬ ▬ ▬ • •	▬ • • • •
9	▬ ▬ ▬ ▬ •	▬ • • ▬
0	▬ ▬ ▬ ▬ ▬	▬▬▬▬
Period (.)	• ▬ • ▬ • ▬	• • ▬ ▬ • •
Comma (,)	▬ ▬ • • ▬ ▬	• ▬ • ▬
Interrogation (?)	• • ▬ ▬ • •	▬ • • ▬ •
Colon (:)	▬ ▬ ▬ • • •	▬ • ▬ • •
Semicolon (;)	▬ • ▬ • ▬ •	• • • • •
Hyphen (-)	▬ • • • • ▬	• • • • • ▬ • •
Slash (/)	▬ • • ▬ •	• • ▬ ▬
Quotation marks (")	• ▬ • • ▬ •	• • ▬ • ▬ •

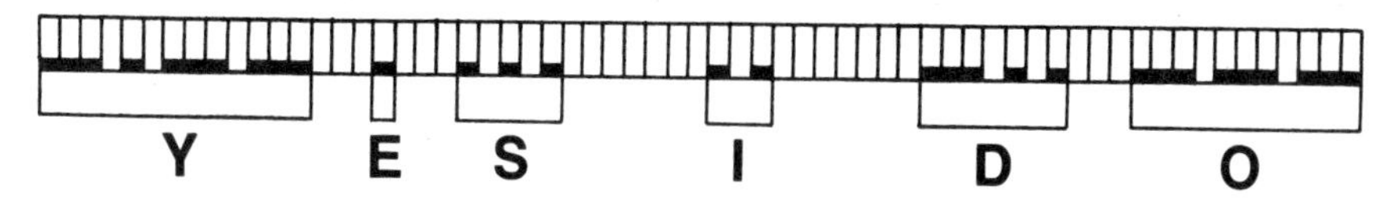

Spacing chart showing the phrase YES I DO

	Duration in no. of dits
Dit	1
Dah	3
Space between elements	1
Space between characters	3
Space between words	7

When comparing the difference between the two codes, notice that the original American Morse contains some spaces between elements much longer than the length of a dit (for example, the letters C and O). Also, some dashes are longer than others (notice the L and zero).

As simple as the code appears, it is interesting to note that Samuel Morse considered its design to be the most difficult part of his invention. Indeed, a great deal of thought must have gone into its conception. To point out just one of its characteristics, notice that the most frequently used characters are the easiest to send and receive. E is the most often used letter, consisting of a single dit. Q and Z are used infrequently, and are examples of some of the more difficult characters to use. Morse determined the frequency of use of each character by visiting a nearby printer. He searched through the printer's type case and counted the number of pieces of type on hand for each character.

CHAPTER FOUR

Learning to Receive and Send

You can at once dispel any notions that you may be "too young," "too old" or "not smart enough" to learn the code. All of these perceptions have been proven wrong.

Many studies about learning the code have been reported in various journals such as *The Psychological Review*, *Psychological Bulletin*, *Journal of Applied Psychology* and *American Journal of Psychology*. All showed little correlation between code aptitude and age or intelligence. Obviously, certain individuals will learn faster than others. But anyone willing to put in a relatively small amount of time and effort can master, and become proficient in, the code. That means you can, too!

Best Way to Learn

These same journals and others have studied various techniques of learning the code. Most are discussed in this chapter, but it is hard to find a substitute for "practice makes perfect."

All individuals do not learn best in the *same way*. Listed below are some of the different techniques and facets of learning that have been suggested. Some may consider such a list presented at the beginning of a chapter on learning the code to be confusing. However, it seems to make sense to present all options initially, rather than have the reader learn at the end that he or she could have benefited by using a different approach.

Various Techniques of Learning

1) **Learn in groups, beginning with the letters comprised of all dits first, going on to letters with all *dahs* next, then learning letters with both dits and *dahs*.** This method makes it somewhat easier to learn initially. Learning the letter for *dah*, then the letter for *dah•dah,* next the letter for *dah•dah•dah* and so on gives one's practice a semblance of organization or grouping and makes it easier to recall certain characters.

2) **Learn the code in groups of letters that have related sounds.** One example would be a group of characters that begin with two dits each, such as U (di•di•*dah*), F (di•di•*dah*•dit) and question mark (di•di•*dah•dah*•di•dit). The argument for this method is similar to that of the one above. It presents characters in groups that contain similar characteristics and therefore make them somewhat easier to remember.

3) **Learn in groups in which each character is not related to another in any way whatsoever.** This technique is based on the fact that characters in real life situations appear in no predictable order, and therefore the student should learn in this manner initially.

4) **Learn the more frequently used letters first and the more difficult ones last.** This means of learning the code is promoted on the argument that the most used letters should be the most familiar to the operator.

5) **Never look at a chart of the code when you practice.** The process of using a chart is discouraged on the grounds that the code is heard and not seen. Therefore, all practice should be of an audio nature, not visual.

6) **Have a chart in front of you at all times.** This technique is promoted on the theory that a chart aids in recollection of the characters and that visual techniques of learning the code are just as productive as auditory techniques.

7) **Listen to characters sent at high speed, with long pauses between each.** This system is based on the belief that an operator is capable of recognizing a *single* Morse character sent at a speed substantially faster than his word or sentence-copying ability. To give an example, if a learner is able to receive at a speed of 5 WPM, the argument is presented that he can copy at this 5 WPM rate just as easily, or easier, if the individual characters are sent at a higher rate of speed (perhaps 10 WPM). Transmitting in this way allows for a longer time period between characters, and as a result the operator has considerably more time to make a decision on what each of the characters are.

There is another advantage to practicing in this manner. By listening to rapidly sent letters and numbers the student becomes accustomed to recognizing the *sound* of a character rather than the combination of dits and *dahs* that comprise it. Also, it is argued that learning the code by using high-speed characters from the beginning of training enables one to advance more quickly to higher rates of code copying.

This type of character presentation is known as the "Farnsworth method," and there is considerable research to prove that the technique is superior. In fact, the ARRL uses the Farnsworth method in its own code-practice material.

The Farnsworth Method

The Farnsworth method of code instruction was named after a blind Amateur Radio operator, Donald R. (Russ) Farnsworth, who during his lifetime alternately held the amateur calls W9SUV, W6TTB and WØJYC. He was first licensed in the mid 1930's and is a Silent Key today. His method of instruction was developed during the late 1950's when Farnsworth asked Bart Bartlett, W6OWP, to help him with the preparation of a Morse code teaching course. A Kleinschmidt tape perforator was used to prepare perforated tapes, capable of sending precisely timed code, which were used as the original masters for his Epsilon Records code course.

Farnsworth's records used code sent at a constant speed of 13 WPM, beginning with simple text which gradually increased in complexity. However, the 13 WPM rate was perfectly sent Morse, using correct element and character spacing. It did not employ the Farnsworth method which bears this man's name today.

The true origin of the Farnsworth method is still a mystery. It is possible that various amateurs used it at different times, each finding it to be a superior method of code instruction, and as a result it eventually became the commonly used technique that it is today.

8) **Use flash cards to learn.** This process is based on the assumption that visual techniques are just as effective or more effective than audio techniques in learning the code.

Interestingly, after the outbreak of World War II, the U.S. Navy, with the help of a renowned psychologist, devised a visual "card game" method of learning the code. This technique used flash cards with an alphabetic or numeric symbol on each. The reverse side of the card contained the Morse code symbol for each character. The object of the game was to guess the Morse equivalent of each character as the cards were displayed.

How Long Will It Take?

No one can answer that question for you individually, but average figures are listed below. WPM stands for words per minute. A word is a group of five characters. The learning time necessary to attain a certain speed is comprised of individual half-hour practice sessions.

Receiving speed attained (WPM)	*Approximate learning time (hours)*
5	30
10	60
15	95
20	150

Do It Right

The best way to learn a language, or anything else for that matter, is to learn it *right* the *first* time. Don't be a "hunt and peck" telegrapher. Study by the following rules.

1) Never study when tired.

2) Don't push yourself. Trying to rush things will cause you to become confused and tired.

3) It is best to study in short sessions, especially in the beginning. Work in 15- to 30-minute periods.

4) Set aside one or two practice periods each day and stick to them.

5) Go by the book. Follow each lesson carefully through to its completion.

6) You may, or may not, reach several "plateaus" as you

progress. Plateaus are pauses in learning speed that sometimes occur at various points in a learning program. Anyone who has taken typing in school is familiar with them. Studies have proven that plateaus are not a universal phenomenon so you may never encounter one. But if you do, relax. Just stick with it and don't become discouraged. You *will* eventually overcome this phantom resistance point and succeed in reaching your code speed goal.

Let's Get Started

The most important things to keep in mind when learning Morse code are:

1) The code is fun!

2) The code is a language, but one much easier to learn than other languages.

3) It is comprised of characters that are made up of "dits" and "dahs."

The proper way to learn code is by sound, because it is easier to "sound out" a character using dits and dahs. The letter Q (– –•–), for example, is easier to say as *dah*•*dah*•di•*dah* than dash•dash•dot•dash.

Q was used to explain why dits and dahs are used, but let's forget about letters for a minute and concentrate on the fundamental elements that make up all characters in the Morse code. These are the dit, the dah, and the space between dits and dahs. Chapter 3 indicated that a dah was three times as long as a dit and that the space between dits and dahs was equal to the length of one dit. Another way of looking at this is to consider the length of a dit as the *fundamental element length* upon which all other element lengths are based. Then we can say that

- The length of a dit = 1 dit length (obviously)
- The length of a space = 1 dit length
- The length of a dah = 3 dit lengths

Begin by pronouncing a string of dits. When you do so, drop the *t* in all dits except the last. This will make the string much easier to pronounce.

di•di•di•di•di•di•di•di•di•dit

Say it fast and say it over and over again. It should sound something like a machine gun burst. (Don't do it in public!)

Next try a string of dahs. Each dah should last three times as long as a dit.

dah•dah•dah•dah•dah•dah•dah•dah•dah•dah

Say this string over and over again also. As you say each string, notice that the individual elements were almost slurred together. This is as it should be. Also, it should have taken about twice as long to say the string of 10 dahs as it did to say the string of 10 dits. That's because there are about twice as many fundamental element lengths (the length of a dit) in a string of 10 dahs as there are in a string of 10 dits. Figure it out. Ten dits with a space between each take up 19 fundamental element lengths (10 dits and 9 spaces). Ten dahs with a space between each take up 39 fundamental element lengths (10 dahs equal 30 fundamental element lengths and 9 spaces equal 9 fundamental element lengths).

Try both exercises several more times to get the feel of it. You are on your way!

Now, back to the letter Q. It should be pronounced *dah•dah*•di•*dah*. You will notice several things about this pronunciation. First, the individual element sounds are combined into a single, multipart word. In combining in this manner, the *t* in the dit is dropped, creating a word that is easier to pronounce. One might say the resulting word is almost melodious. Taking matters a step further, certain parts of the word (the dahs) are emphasized. The emphasized portions are indicated by *italic type*. Now say the Q out loud, making sure you pronounce it correctly and use the proper emphasis.

dah•dah•di•*dah*

Let's try another character, the error indicator. If the sender makes a mistake, he indicates this by sending a string of eight dits. This character is pronounced

di•di•di•di•di•di•di•dit

The *t* in all dits except the last are dropped, making it easier to pronounce.

When saying each character you must also keep in mind the element length and spacing. This may seem like a lot to remember, and it is when you first start out. Eventually, it will become second nature. Here are the rules for sounding out each character. Read them, then say the Q and error indicator again.

1) Combine individual element sounds into a single word.

2) When combining, drop the *t* in all dits except one that falls at the end of a word.

3) Accentuate the dahs.

4) Space the sounding out of the dits and dahs so that:

A—A dah is three times as long as a dit.

B—The time lag between elements in a single character is equal to the length of a dit.

Now pronounce the Q and error indicator once more.

dah•*dah*•di•*dah*

di•di•di•di•di•di•di•dit

Both of these characters were introduced in the beginning so you could master the basic rules for learning code. Consider them to be sample exercises.

We shall now begin our actual study of the 45 characters comprising the Morse code. These are best learned in groups. The method that shall be used is that of presenting the characters in groups of related sounds. There is, however, another section following the groupings by related sounds that represents characters in no related order whatsoever. This is provided for students or instructors who feel that this alternative method may be superior.

Group 1

E	dit
T	*dah*
A	di•*dah*
R	di•*dah*•dit
Period (.)	di•*dah*•di•*dah*•di•*dah*

Study the first group carefully. Say each letter to yourself several times. Use the rules of pronunciation you have learned. Truncate all dits except those that fall at the end. Accentuate the dahs. Remember your rules for spacing. Say these characters over again and again. Memorize them. They are an easy group to learn. Do not go past this point unless you know them cold.

After they are learned, it would be helpful if a friend could assist by calling out the character sounds while you write down the characters they represent. Anyone can do this. Knowledge of the code is not required as long as the characters are pronounced correctly. Of course, if someone who knows the code is available, so much the better.

Some texts recommend that the characters be printed as they are copied; others suggest writing them in longhand. It's best to start your copying practice by printing. Written letters are sometimes confused with one another. This can be especially troublesome when copying meaningless groups of letters.

Once you have mastered Group 1, continue to the next group.

Group 2

N	*dah*•dit
D	*dah*•di•dit
K	*dah*•di•*dah*
C	*dah*•di•*dah*•dit
X	*dah*•di•di•*dah*
Y	*dah*•di•*dah*•*dah*

Study this group in the same manner, again using a friend if possible. Once you have mastered this group, go back and review Group 1. Then, have your "instructor" sound out the characters from both groups combined, jumping back and forth between groups. Sample learning exercises are provided immediately following this subchapter. They are designed so that an exercise for any group includes characters from all groups before it. This way your practice will be cumulative and you will not forget what you already learned. Your instructor (or you yourself) should use these exercises to help you practice immediately after you have memorized each group.

If you are not sure of a character as you listen to the code, skip it. Don't look back. It is better to miss one character than two or three following it. Once you have mastered Groups 1 and 2 and have done the sample exercises, you may continue.

Group 3

I	di•dit
S	di•di•dit
H	di•di•di•dit
V	di•di•di•*dah*
J	di•*dah•dah•dah*
B	*dah*•di•di•dit

Group 3 is relatively easy to learn. Use the same techniques as you did for the first groups. Once you learn this set of characters, have your instructor intermix characters from all three groups. He or she may even make up words from these groups. Remember, the correct spacing between words is equal to seven dits. See Sample Learning Exercise 2. When you copy words, it is important that you *don't anticipate* what letters are to come next. The obvious reason for this is that you may anticipate incorrectly. Once you have mastered all characters to this point, you should proceed with Group 4.

Group 4

W	di•*dah•dah*
L	di•*dah*•di•dit
P	di•*dah•dah*•dit
U	di•di•*dah*
F	di•di•*dah*•dit
Question Mark (?)	di•di•*dah•dah*•di•dit

Your learning exercises will be getting progressively more challenging as your vocabulary becomes larger. Don't rush matters; learn to enjoy the code. Here's another group.

Group 5

M	*dah•dah*
O	*dah•dah•dah*
G	*dah•dah*•dit
Z	*dah•dah*•di•dit
Q	*dah•dah*•di•*dah*
Comma (,)	*dah•dah*•di•di•*dah•dah*

Once you have mastered Group 5, you will be familiar with all letters of the alphabet and the most-used punctuation marks. That's quite an accomplishment! You have a lot to remember at this point, so don't be discouraged if things seem to move slowly. Stick to your scheduled practice sessions and take it one step at a time.

After Sample Learning Exercise 4 has been completed, you will be in a position to do something that is very helpful in practicing the code. As you go about your daily routine, convert words you see into code and say them to yourself in Morse. Do this with street signs, posters, billboards, and so forth. Think *code*. You will be surprised how quickly your speed improves.

There are only a few more groups to come; you are near the end. Numbers are next. They are easy to remember.

Group 6

1	di•*dah•dah•dah•dah*
2	di•di•*dah•dah•dah*
3	di•di•di•*dah•dah*
4	di•di•di•di•*dah*
5	di•di•di•di•dit

Never practice more than an hour or two each day and take frequent breaks. Learning code requires your undivided attention. Study in short intervals with rest periods between.

Here's the last group of numbers. It includes the error indicator that was used as an example earlier.

Group 7

6	*dah*•di•di•di•dit
7	*dah•dah*•di•di•dit
8	*dah•dah•dah*•di•dit
9	*dah•dah•dah•dah*•dit
0	*dah•dah•dah•dah•dah*
Error	di•di•di•di•di•di•di•dit

You're practically there now. The next to the last group contains five additional punctuation marks. You will find in actual practice that only the slash (/) is used to any great extent. Depending upon the types of stations you listen to, it is possible that the infrequency of use of the others may make them difficult to recall.

Group 8

Quotation (")	di•*dah*•di•di•*dah*•dit
Semicolon (;)	*dah*•di•*dah*•di•*dah*•dit
Slash (/)	*dah*•di•di•*dah*•dit
Hyphen (-)	*dah*•di•di•di•di•*dah*
Colon (:)	*dah*•*dah*•*dah*•di•di•dit

Finally, the last group, consisting of special symbols that are used during telegraphy conversations to expedite transmission. Seven of the symbols are actually combinations of letters sent as a *single character*.

Group 9

Message received OK (R)	di•*dah*•dit
Invitation to transmit (K) to any station	*dah*•di•*dah*
Invitation to transmit ($\overline{KN}$) only to station in contact with	*dah*•di•*dah*•*dah*•dit
Wait ($\overline{AS}$)	di•*dah*•di•di•dit
Break ($\overline{BT}$)	*dah*•di•di•di•*dah*
End of message ($\overline{AR}$)	di•*dah*•di•*dah*•dit
End of contact ($\overline{SK}$)	di•di•di•*dah*•di•*dah*
Understood ($\overline{SN}$)	di•di•di•*dah*•dit
Attention ($\overline{KA}$)	*dah*•di•*dah*•di•*dah*

Congratulations! You have learned the entire Morse code. There is a good chance you have begun to use the code already, perhaps by listening to shortwave broadcasts. If so, that's great. You will want to improve your skills, and the section on Code Practice Methods is designed to help you attain this goal.

International Morse Code

A	di•*dah*	•—
B	*dah*•di•di•dit	—•••
C	*dah*•di•*dah*•dit	—•—•
D	*dah*•di•dit	—••
E	dit	•
F	di•di•*dah*•dit	••—•
G	*dah*•*dah*•dit	——•
H	di•di•di•dit	••••
I	di•dit	••
J	di•*dah*•*dah*•*dah*	•———
K	*dah*•di•*dah*	—•—
L	di•*dah*•di•dit	•—••
M	*dah*•*dah*	——
N	*dah*•dit	—•
O	*dah*•*dah*•*dah*	———
P	di•*dah*•*dah*•dit	•——•
Q	*dah*•*dah*•di•*dah*	——•—
R	di•*dah*•dit	•—•
S	di•di•dit	•••
T	*dah*	—
U	di•di•*dah*	••—
V	di•di•di•*dah*	•••—
W	di•*dah*•*dah*	•——
X	*dah*•di•di•*dah*	—••—
Y	*dah*•di•*dah*•*dah*	—•——
Z	*dah*•*dah*•di•dit	——••
1	di•*dah*•*dah*•*dah*•*dah*	•————
2	di•di•*dah*•*dah*•*dah*	••———
3	di•di•di•*dah*•*dah*	•••——
4	di•di•di•di•*dah*	••••—
5	di•di•di•di•dit	•••••
6	*dah*•di•di•di•dit	—••••
7	*dah*•*dah*•di•di•dit	——•••
8	*dah*•*dah*•*dah*•di•dit	———••
9	*dah*•*dah*•*dah*•*dah*•dit	————•
0	*dah*•*dah*•*dah*•*dah*•*dah*	—————

Period (.)	di•*dah*•di•*dah*•di•*dah*	•—•—•—
Comma (,)	*dah*•*dah*•di•di•*dah*•*dah*	——••——
Question Mark (?)	di•di•*dah*•*dah*•di•dit	••——••
Colon (:)	*dah*•*dah*•*dah*•di•di•dit	———•••
Semicolon (;)	*dah*•di•*dah*•di•*dah*•dit	—•—•—•
Hyphen (-)	*dah*•di•di•di•di•*dah*	—••••—
Slash (/)	*dah*•di•di•*dah*•dit	—••—•
Quotation ('')	di•*dah*•di•di•*dah*•dit	•—••—•
Error	di•di•di•di•di•di•di•dit	••••••••

Additional, Lesser-Used Morse Characters

Apostrophe	[']	di•*dah*•*dah*•*dah*•*dah*•dit	•————•	$\overline{WG}$
Left-hand bracket (parenthesis)	[(]	*dah*•di•*dah*•*dah*•dit	—•——•	$\overline{KN}$
Right-hand bracket (parenthesis)	[)]	*dah*•di•*dah*•*dah*•di•*dah*	—•——•—	$\overline{KK}$
Double hyphen	[=]	*dah*•di•di•di•*dah*	—•••—	$\overline{BT}$
Cross or addition sign	[+]	di•*dah*•di•*dah*•dit	•—•—•	$\overline{AR}$
Multiplication sign	[×]	*dah*•di•di•*dah*	—••—	$\overline{NA}$
Dollar sign	[$]	di•di•di•*dah*•di•di•*dah*	•••—••—	$\overline{SX}$
Paragraph	[¶]	di•*dah*•di•*dah*•di•dit	•—•—••	$\overline{AL}$
Underline	[__]	di•di•*dah*•*dah*•di•*dah*	••——•—	$\overline{IQ}$

The following letters are used in certain European languages which use the Latin alphabet.

Ä, Ą	didahdidah	•—•—
Á, Å, À, Â	didahdahdidah	•——•—
Ç, Ć	dahdidahdidit	—•—••
É, È, Ę	dididahdidit	••—••
È	didahdididah	•—••—
Ê	dahdididahdit	—••—•
Ö, O̊, Ó	dahdahdahdit	———•
Ñ	dahdahdidahdah	——•——
Ü	dididahdah	••——
Ż	dahdahdidit	——••
Ź	dahdahdididah	——••—
CH, Ş	dahdahdahdah	————

Special Esperanto characters

Ĉ	dahdidahdidit	—•—••
Ŝ	dididididahdit	•••—•
Ĵ	didahdahdahdit	•———•
Ĥ	dahdidahdahdit	—•——•
Û	dididahdah	••——

Signals used in other radio services

Interrogatory	dididahdidah	••—•—	$\overline{\text{INT}}$
Emergency silence	didididIdahdah	••••——	$\overline{\text{HM}}$
Executive follows	dididahdididah	••—••—	$\overline{\text{IX}}$
Break-in signal	dahdahdahdahdah	—————	$\overline{\text{TTTTT}}$
Emergency signal	didididahdahdahdididit	•••———•••	$\overline{\text{SOS}}$
Relay of distress	dahdididahdididahdidit	—••—••—••	$\overline{\text{DDD}}$

Abbreviated Numbers

Where there can be no misunderstanding as to the meaning, numbers may be represented in an abbreviated form as follows.

1	di•*dah*	•—	(A)
2	di•di•*dah*	••—	(U)
3	di•di•di•*dah*	•••—	(V)
4	di•di•di•di•*dah*	••••—	(4)
5	di•di•di•di•dit	••••• or •	(5 or E)
6	*dah*•di•di•di•dit	—••••	(6)
7	*dah*•di•di•dit	—•••	(B)
8	*dah*•di•dit	—••	(D)
9	*dah*•dit	—•	(N)
Ø	*dah*	—	(T)

Fractions

To transmit a fractional number, transmit a slash between the numeric components of the fraction, as follows.

Two thirds is transmitted as 2/3.

To transmit a number that includes a fraction, transmit a hyphen between the whole number component and the fraction itself.

7 1/8 is transmitted as 7-1/8.

Percentages

To indicate the percentage sign (%), transmit the figure zero (*dah•dah•dah•dah•dah*) followed by the slash (*dah*•di•di•*dah*•dit) and the figure zero again (*dah•dah•dah•dah•dah*) (Ø/Ø).

To transmit a whole number followed by a percentage sign, a fractional number followed by a percentage sign or a fraction followed by a percentage sign, transmit a single hyphen between each component of the number, as follows.

3% is transmitted as 3-Ø/Ø

4-1/2% is transmitted as 4-1/2-Ø/Ø

1/5% is transmitted as 1/5-Ø/Ø

Minutes and Seconds

To transmit the minute sign (′) or second sign (″) as used, for example, in longitudinal or latitudinal coordinates (5′ 23″), use the apostrophe (di•*dah•dah•dah•dah*•dit) once or twice as appropriate. Do not use the quotation symbol.

Sample Learning Exercises

Exercise 1 (for Groups 1 and 2)

DXETN	X.ADE	RKCA.
KCATR	RNYXT	XARCN
CYTA.	TCRYE	Y.RKC
RNCAK	DYNTE	DXA.C
DKC.A	R.CTY	RKXEC
XAN.C	TNXAC	YTAKN
E.KAC	NTRXE	EXRYN
ADTY.	NY.TD	RYACT
E.XYA	TDA.D	NYEDT
KY.TR	C.TXD	CAKEX
XADRY	DTCAX	DKEAC
YKERC	T.XRY	YDDTN
ERNKC	RYACR	AKEX.
EYACR	NXTE.	DAYEC
XKNR.	TRCDA	KXACN
YE.KC	AYNTX	DT.XN
KAEDR	RN.TY	XEDAC
YCNT.	DCTRA	KEYNR
CRAXD	KDTE.	XNYR.
DYEXN	KARXD	NKATY
CAR	TARDY	TAR
END.	EXTEND	YARN
KEN	DENT	TAX

TAN	RAT	TEN
XRAY	EXTRA	TARDY
ANNEX	YEAR.	AXE
CAT	DARN	RACK
RANDY	TAKE	CARD.
RACKET	RAN	TACK
CAN	ACE	DEAR
EAT	ADD.	RANK
EAR	RARE	TACKY
CANT	TANK	CANDY
ACCEDE	CART	EARN
AND	END	ACT
TRACK	READ.	TREAD
DANCE	ANDY	TREAT
EDDY	ACCENT	READY
EARNED	REED	ADDED
RED	ENDED	DECK

RANDY EARNED A CANDY TREAT.

KEN CANT DANCE.

A KEY CAN END A TARDY TANK.

A DEER CANT EAT A DEAD TREE.

AN EXTRA RAT DANCED DANDY.

TEN DEER CAN READ AN EXTENDED CARD.

DAN TAXED AN AXE AND A RED CAT ENTERED A CAR DENT.

Exercise 2 (for Groups 1, 2 and 3)

VAJKB	V.KEJ	IX.HD
JDEVY	SX.VC	XRHNE
JXABE	DEJNT	SAICY
NRHKT	TVCRI	B.XAH
RBDTI	SDEBY	VC.IX
BKESK	BT.KI	IC.VN
XEJC.	VCRSD	KAJNT
AVDTJ	RSDTH	THYEV
SNEJX	ABCTV	IDTSN
TIKAJ	BCRSN	CRHNE
DINTB	N.VJK	IYTSX
KAJYE	RBKES	NRHYT
TJK.V	IXAHC	S.VXA
CATJV	XBR.D	YESVH

KJTIA	YNEV.	BSDNR
CIHTJ	IEVJN	C.RAI
HBRXE	CDSVI	YAX.E
RDNHI	VJKTA	HKXJD
YRIHV	JEASD	H.NXJ
ITSVX	BNYCK	HRJVD
HIS	EAT	RAIN
DEAR	VANISH	YANK
SIT.	VACATE	JEER
XRAY	SIXTH	KEEN
JAR	XAVIER	DEAD
BEAR	HIS	YEAR
CATER	KICK.	JIST
BAT	VAIN	TEAR.
CHARR	JINX	JAB
ICESKATE	VERY	HAT
SET	HASH	BEARD
IDEA	JAR	BEAK
INCASE	BREAK	HASTY
SANDY	BEND.	HARSH
SAD	SAVED	BEATEN
HACK	INCH	BANDS
BETTY	JERSEY	HANNA
THE	IS	HAS
SEEDS	BAR.	SECT
SEA	SIEVE	HARDEN

JACK SET A BAT IN A JAR.

BETTY SAVED THE HASTY DEER IN JERSEY.

THE BEAR BENDS SEEDS IN THE SEA.

SIX BATS HAVE VERY HARD HEADS.

A CAB IN JERSEY HAS INACTIVE SEATS.

HIS VERY BEATEN HATBAND INCITES THE CAB DRIVER.

JACK STAYS NEAR IN CASE HANNA BREAKS THE ICESKATE.

Exercise 3 (for Groups 1, 2, 3 and 4)

UFVCA	DRPHT	YAFIC
PSXAF	IYN.R	W?SXR
FSY.W	E?HD.	WID.?
UPHJK	PFVKE	VNE?J
ULVXT	VNELJ	XERLP

LJY.T	CTUID	WJKRF
KRFBY	BIDXT	?HNAW
APBX.	BXAPL	A?HNT
F.I?S	BDTFH	WFHR?
IY.?H	PVCEU	DKRAU
X.E?J	LVNEU	NEUVC
UFHVX	WSHRA	KALUI
SXTPJ	RLINT	TAW?B
CXFTY	?CASJ	LDN.A
YTASW	PXKEB	JED?C
XSRWU	AVUHX	CFREL
D.NSW	INLFD	ICA.S
HDUVH	YEPWK	NJ?IC
U.VXR	TYBPX	CWFDN
KESVF	H.XCD	AWLRY
WELL	WIDEN	EXTRA
KINDLE	PART	UP?
UNDER	FIXED	JINX
FAT	DUN?	FRILL
PUNCH	WISH	YEARS
RUN.	FINAL	JET
WIND	PINCH	WASH
WINCH	KINDEST	PARK.
VULCHER	WAIST	WILL
LAND	UNADVISED	PAT
WARN	FAIL	WISH
LEAN	UNDER	PATTERN
ULCER	PANDA	UNCERTAIN
LETTER	FRANK	PARTY
UNBIND	PICK	FADE?
ULTRA	LATIN	FRAN
LIVED.	PET	FEATHER
PETTY	LAST	PEAK
FEAT	FEET?	PATIENT
UNAWARE	PEAL	UNACCENTED

DID LARRY PUSH THE LANTERN UNDER THE CART?

LYNN ADVISED PAT IN LATIN THAT THE FACULTY FAILED WHEN THEY LEARNED THE PANDA LEFT.

WAS THE UKULELE UNBALANCED WHEN IT WAS LEANED FARTHER INWARD?

WHEN DID WILLY WANT FEATHERS IN HIS UNFEATHERED HAT?

DIDNT LINDA PUSH THE SEAL IN THE WATER WHEN SHE LEARNED THAT IT WASNT WET?

WILL FANNY WISH HER FEATHERED FRIENDS EAT ALL THE BIRDSEED LEFT BY WANDA?

WHY DIDNT HARRY PICK UP THE FALLEN FAUCET WHEN IT FELL UNDER THE SINK?

Exercise 4 (for Groups 1, 2, 3, 4 and 5)

QGNBC
DHOCE
DGMYW
UODHO
IUOCE
ITQHP
GIRIA
TNABD
CFLWT
JSLVX
FLXVR
NABDG
HOCEJ
MZYWS
RI.IR
IRVTM
NBCEJ
SLXVQ
KVRIQ
JSKVQ
MACHINE
ZANY
ZIP.
MAD
GALLON
ORDEAL
OAR
QUACK
QUANTITY
MAGIC.
OBJECT

EIRJT
JGNBA
LJSWS
CEJSR
LSJXV
NBDHP
CFLXU
LWTON
LWSKU
QHOCE
IJSKU
PEIRJ
BCFLW
OCFKV
IUHPF
PEJSK
TMZYX
SKVR.
QDBGN
TNLWA
OAK
QUAIL
OPOSSUM
GEAR
MUNCH?
QUIRK
ZEALOT,
QUARRY
OIL
QUEEN
OBEDIENT

VPFLX
HPEIQ
OGMZY
YXUOD
KVRJS
VRJTM
UGMYX
XUODG
GMYCW
LXVRI
ZYWTM
OCEIQ
MYXVR
PMZYI
R,TMY
Z?YGI
GMYWT
?TNAB
XUODH
WSKVR
GABLE
MEND
GENERAL
ODOMETER
GAME,
MADE
ZESTFUL
ZIGZAG
ZENITH
GADGET
GLASS

ZEAL.
QUICK
GAGE
QUAKER
GERMANY
MUSCLE,
OLIVE
GEARSHIFT
QUAD

MARY?
MENTAL
ZEPHYR
GLOW,
MUMBLE
ZEBRA
ZEPPELIN
ZINC.
ZERO?

GHOST
OFFSET
MENU
MOLTEN
QUESTION?
OCTOBER
ORANGE
QUIET
QUARTER

THE GEARS IN THE MACHINE WORKED, BUT THE GALLON PER MILE CONSUMPTION WAS QUITE HIGH.

WAS THE ZEPPELIN QUIET WHEN IT TOOK OFF, OR DID THE ENGINES MAKE A LOUD NOISE?

THE QUICKEST WAY FOR A QUAIL TO ESCAPE A ZEALOUS HUNTER IS TO FLY IN A ZIGZAG FASHION.

I BELIEVE AN OPOSSUM CAN GO WHERE ZEBRAS CANT, ISNT THAT SO?

DID THE MOUNTAIN CLIMBER REACH THE SUMMIT QUICKLY, OR DID THE BLOWING SAND PRESENT A PROBLEM?

ZEROING IN ON A QUACKING DUCK IS EASY TO DO IF IT ISNT ZIPPING ALONG AT A RAPID GATE.

REACHING THE ZENITH IS QUITE A CHALLENGE IF THE ORDEAL IS WORSENED BY QUIRKS OF NATURE.

Exercise 5 (for Groups 1, 2, 3, 4, 5 and 6)

,1ZLX
ZLWEI
3YJS.
JTF?2
GNLP5
RP4UA
Q1YIR
2YIRY
OMZKV
WFKVD
TICFK
OT?EW
2RJ3R

4VDHO
VFCKV
LMLTY
,4RIE
XGNUH
2QEYI
E3ROI
?TWVU
3S,RL
3S.OH
R14UB
T?SNM
5JPFX

IREL4
,5YIQ
P4AGM
14.VC
NLO3T
UBCEI
4GP2V
YIR14
R1LTV
?HVY.
Q.YIQ
CEJTX
VCEIQ

DGN3E	4RIYA	A,45V
S?PHW	X5NGY	13TX.
VDGN3	HDO3T	,LWES
,ASFW	X.Y41	Y.2RO
?XV.,	OH3TF	3YIFE
LO4TX	P4UBD	GNLON
LXHOA	LMYST	U2REA
MIQEY	?,2YN	JT,UT
S.R3X	SCPF4	3S?2V
12RIP	,4LO3	YIA3I
,EYJT	HO3S,	2Q.YI
Q4ZLX	?X,LZ	,2PRT
IR41X	ON12R	23T?S
HP4VC	KVDD,	KVCFK
BDGMZ	S.4SC	P4VD,
V.24R	5ZLXG	Q4PTM
GNLP5	LXHO3	,5YIQ
HO3TC	RL5UB	5URLP
KFUAB	HO2QL	UCESV
KUBDX	CEJS,	P5WEI
J4UAA	MYJT?	S.23V
XGMYI	T?5,B	TOSAF
5.2RL	DHP5W	2QLEM
D1P5V	LO3CT	CEJS,
MYIQ,	2WSAI	QZLMY
BCEJS	X3BNH	DHP5X
Q,2P5	ANRIP	BLM4C

IT TOOK 1,235 MINUTES TO GO 41 MILES IN 5 DAYS WHEN PULLING A TRAIN OF 34 CARS WITH 12 CONDUCTORS.

51 MINUTES AFTER 22 MEN RAN 13 MILES, EACH MAN DRANK 3 GLASSES OF WATER IN 32 MINUTES, 5 HOURS BEFORE THEY RAN 11 MORE MILES IN 132 MINUTES.

111 TADPOLES WERE EATEN BY 23 FROGS WHICH WERE EATEN BY 5 FISH WHICH WERE CAUGHT BY 2 FISHERMAN.

AN AVERAGE OF 21,135 MOSQUITOES OCCUPY AN AREA OF 13 ACRES OVER A PERIOD OF 23 DAYS BETWEEN THE 21ST OF JULY AND THE 30TH OF AUGUST.

THERE ARE 123,541,342 ATOMS IN EACH 47 CUBIC MILES OF ATMOSPHERE ABOVE A HEIGHT OF 13,221 METERS IN ANY GIVEN 21 HOUR PERIOD.

IT IS ESTIMATED THAT 31,542,111,234 GRAINS OF SAND ARE IN AN 11 CUBIC MILE AREA OF THE SAHARA DESERT WHICH IS LOCATED AT 24 DEGREES 15 MINUTES NORTH, AND 13 DEGREES 21 MINUTES WEST.

Exercise 6 (for Groups 1, 2, 3, 4, 5, 6 and 7)

XGMZK	3S,6Z	XHP4V
BCF3U	0SMYJ	LO2Q6
A4U,A	3S.Z9	2R71Q
P4UA1	KVCEJ	O2Q7Z
K?FCV	84UBD	M39,A
VDHO3	WETLG	DGMZK
8CLB.	S.95W	ZLXHO
8BDHE	TXHO3	6YJT?
BDHP5	S,7M0	LXH5P
RI95X	OH3S.	PERKO
T,.94	?U4OP	UASFP
EIY6T	EIR7L	2R95S
P4VDG	TY4,A	,74HW
XGMZL	YIR3O	XGMYI
GN0NL	7S,6T	3S.YU
VCKUF	IR87T	ABCFL
JS,1M	P4UBC	,6ZKV
MZLXH	9.IVD	85XGM
WERI8	BUAGD	Q7NP5
P5WFL	ZKAHB	WFKUA
ABDGN	V.36S	WFKVD
O2R83	EJS.A	DOH3S
O3TB7	NLP5W	KVDHP
AS.59	M.YIQ	WEOQ6
WE.JT	CBFKU	DARGM
0NLO2	7.5T,	HP4VC
TE.71	2R83S	U,6YM

5XHP1	ERQI8	4UACB
VCFLW	IYR86	ZLWFL
,6ZLW	DGMYJ	QLU6Y
Q71MY	2JBN6	LKVUJ
O3TN?	7.4Q6	T?7VK
XGMYJ	Q6ZLW	EIQ70
EIN4P	R59XG	WEJS,
KVFEC	T58YR	LTX,7
JTV.8	YRI95	KUAEQ
,6YIQ	YJS,0	95XHP
T?D.8	EJS.8	MZLXH
UABCF	MZXHL	1DP5W
T?J,Q	T?J.U	3O6TB

STATISTICS SHOW THAT 89,170,396 BOOKS ARE IN 6,200 LIBRARIES WITHIN 5,198,032 SQUARE MILES.

APPROXIMATELY 8,546,789,001 STARS OCCUPY 84,560 GALAXIES WITHIN A RADIUS OF 912,314 LIGHT YEARS.

SINCE ITS OPENING, THE FRANKLIN MINT HAS TURNED OUT 9,137,862,450 PENNIES, 7,967,321,012 NICKELS, 6,448,987,304 DIMES, 6,678,013,415 QUARTERS AND 970,862,713 HALF DOLLARS.

THE MOST EXPENSIVE BUILDING EVER BUILT IN PHILADELPHIA COST 680,934,551 DOLLARS. THE SECOND MOST EXPENSIVE WAS 400,796,089 DOLLARS, AND THE THIRD MOST EXPENSIVE WAS 112,986,079 DOLLARS.

144,918,045 SHARES WERE TRADED ON THE N.Y. STOCK EXCHANGE ON MONDAY, 237,662,101 ON TUESDAY, 101,644,983 ON WEDNESDAY, 286,911,005 ON THURSDAY AND 210,099,847 ON FRIDAY.

ONE QUARTER OF THE RADIOS HAVE SERIAL NUMBERS A1927432 THROUGH CB910432, ANOTHER QUARTER D6430019 THROUGH H1036647, ANOTHER QUARTER J7786043 THROUGH M2694003 AND ANOTHER QUARTER P2633498 THROUGH S9887722.

SOCIAL SECURITY NUMBERS 080287001 THROUGH 256301148 WERE ISSUED FROM 1949 TO 1956, 324899706 THROUGH

410086755 FROM 1958 TO 1963 AND 449007632 THROUGH 533864190 FROM 1965 TO 1971.

Exercise 7 (for Groups 1, 2, 3, 4, 5, 6, 7 and 8)

S.9V4
/81P3
T:BY“
?N140
HP4YA
DMGJY
O2HR9
BJX/.
O3S.8
BGAEI
BSC8F
5XIL3
7WLFX
HQ7DM
R8FS.
L3UT/
T?.YR
HP4UA
ZLWFK
95IEW
1.4CV
5WPG6
BCEIQ
UBDHO
R3JR6
KCFVE
WFLWE
6ZLXH
3SI4.
5XGYM
JS.9L
IR8TM
O3.8S
UB9,H
J,TW2

ACK0F
7LMZK
3S,7K
R8HQL
5XSO3
S,6KZ
VCERI
NONLP
I6YR3
S.8:S
U:CBF
9D1AU
02QYJ
T/Y7W
,7LWY
LCVUK
N:AOD
S6.8F
MZLXH
IQ7IO
DNHOR
2,R94
L?F6Z
O2R1N
2E7;N
2Q7XO
AY.CD
Q-O2H
R8M3T
OMZLX
83REC
DO5QY
71NLE
9.-5U
PH5WF

ET8I2
EQ6ZI
KVDNG
2RF8,
KVDHP
0QUN7
VKFCL
0MIYR
7OMYJ
4CVEJ
M8“WE
W1LER
9AB4C
T?0/X
S.93X
SNJ7;
C5?TJ
LF,WF
6YJ,7
4HBUP
LXGN-
/C,B6
WVKND
ZLXHP
CEJT?
O2?7Y
Y-JTX
MO2R8
5XGN0
/.83T
JTX.9
8.L,S
9RQ5A
NLP4X
:WIOL

O3HUB	74BUC	5WETJ
;O8CQ	1FPW5	95JWE
4VDOH	3S;Z	7CS13
2Q6IY	HABNO	GMZLR
YJS.9	"GM:Z	10SMN

"1/3-2/3 OF ALL AMERICANS EAT VEGETABLES AT EACH MEAL" LARRY SAID; "1/2-3/4 EAT MEAT."

THE FOLLOWING IS THE PROPER SERIES OF NUMBERS: 9-8-4-0-0-2-3.

DAN SAID "THIS IS THE COMBINATION TO THE SAFE: 31 LEFT/62 RIGHT/10 LEFT."

"AIRPLANES GO FASTER THAN TRAINS; TRAINS GO FASTER THAN CARS; CARS GO FASTER THAN BIKES" HE SAID.

LOOK FOR NUMBERS 1-11, 110-120, 145-211, 615-819 AND 912-986.

"3/3 IS GREATER THAN 2/3; 2/3 IS GREATER THAN 1/3; 1/3 IS GREATER THAN 1/8" SHE SAID.

THE PROPER MODEL NUMBER IS 2-9-4/38-29; THE PROPER SERIAL NUMBER IS 1-6-7/46/927-4.

Alternative Method of Learning Code by Groups of Characters in no Related Order

Groups of characters should be learned in the order presented here. Complete the practice exercises following each group before going onto the next group of characters. Each exercise includes characters from the preceding groups so that the learning process is cumulative.

Group 1

T *dah*
R di•*dah*•dit
N *dah*•dit
E dit
A di•*dah*

NARET	RTAEN	TAENR
ERTNA	ARNTE	TNAER

RENTA	NRTAE	ENART
ANTRE	TRENA	RNETA
NERTA	ETANR	AETNR
TENRA	RANET	NTREA
EARNT	ATENR	TNERA
NRATE	ERTAN	ARETN
TNAER	RERAT	NAATE
ERTEA	AENRT	TNRAE
RTNEA	NATRE	ETRAN
ATRNE	TNERA	REATR
AT	ARE	ANT
ART	ATE	AN
EARN	EAR	ENTER
EAT	NATE	NEAR
RENT	REAR	RAT
RAN	RARE	RANT
RATE	RETREAT	TART
TEA	TEAR	TAN
TAR	TREAT	TEEN
TREE	TEN	NEAT

NATE ATE A TART

ENTER AT REAR

RENT A RAT

RARE TREAT NEAR TEA TREE

EAT AN ANT EAR

NATE RAN AT A RAT

A RARE RAT RAN AT A NEAT TREE

Group 2

I di•dit
O *dah•dah•dah*
S di•di•dit
D *dah*•di•dit
H di•di•di•dit
C *dah*•di•*dah*•dit

IRSNO OEDCA SHREI

DAOCN	HSNIT	CORHA
TDCAS	INDCR	OERDS
SNICT	DSETO	HCARS
CDIEN	IDATH	OENHC
NRSOI	ADHNC	DEICT
RDSCA	TDSIH	DNSCO
CDREI	SOAEC	CNEID
DAHRE	HESOR	RSCHN
THCIS	NDSCO	CIETN
NDEOC	HIHDA	EHSNR
ONEOH	ODETA	NRHOI
EDSNR	AICST	THOSC
NSCOI	DNRSO	RCADS
HNSRI	EIDCH	NDCTA
DCNIT	INHAO	SCRSI
CHNED	DSETC	EOCHR
AOHDC	NHRAO	ROSHI
ISDTA	NDICR	DCEOI
HONDS	EICNR	HSNIO
DNAOI	SEHCT	ROHDA
CTRSI	HNROA	CRSID
INADH	ORDTC	SNHDI
ADDITION	CADDIE	DANISH
CEASE	CARRIED	DISASTER
DOT	DASH	DOOR
DEAN	ERASER	ENTIRE
HEART	HEARD	HERS
HAIR	HOOD	HOSE
HORROR	HEAR	HATE
COANCHOR	DIET	DICE
HIDDEN	CHIC	DAD
IN	NOT	IS
NATASHA	NED	NOTE
NINETEEN	NOISE	NATION
ONE	RADIO	RETORT
RETIRED	RATE	REASON
RAIN	SHARE	SHRED
STRAIN	SHIRT	STARTED
STOOD	SHEAR	STRANDED

STATION	TINT	TEARS
TEASE	TENTS	THERE
TORNADO	THREE	THIRTEEN
AND	CONED	HIDE
COSINE	HERD	READ
SHED	SHE	STAND

RITA HEARD A TRAIN THEN RAN AND HID

NATASHA SHARES TEA AND TARTS

NATTIE HATES TORNADOES

THE DEER HAD A NOTION TO DICTATE DOTS AND DASHES ON THE RADIO

THE RAT DARTS IN HORROR IN THIRTEEN SEAS

THE STANDARD HIRE RATE TRENDS ON THE THIN SIDE

THE SODA STAINED THE ENTIRE STATE

Group 3

U di•di•*dah*
Y *dah*•di•*dah*•*dah*
. di•*dah*•di•*dah*•di•*dah*
L di•*dah*•di•dit
M *dah*•*dah*
P di•*dah*•*dah*•dit
G *dah*•*dah*•dit

UMTSD	YIL.H	M.NAS
LRPGT	MTU.S	PYNOA
GURNH	URGOD	Y.RHT
.PNSO	LYDEG	MNEAY
PLTDA	GYTNR	UE.SC
YMTIS	L.OHA	CMTGO
OYGNI	RPM.U	NGYSE
T.GYE	SMRAG	EML.U
DYMNR	UL.TI	YPSAH
GPT.1	ALGDS	RMUYN
ELUTD	NYLPO	TYGSH
IU.RO	SOMGT	OPYUH
SH.GP	DLG.U	HMYDS

CLGNR GESYC PNT.E
MUOID LTNPH .UTOG
YIEPL UNSY. ALYDC
EMUNR RY.SC NMUCH
TL.RA IYMNO SPUCD
HUGTN CLOYT UREHD
P.TAH GYSED PTHAE
MNRAU L.TIA OYSEC
SU.GN DMN.O HMLRY
DMUTA RALGH T.MSN
ARMY ALCOHOL CHEAPLY
CAMERA CANADA COMEDIAN
COMMUNICATE CANCEL CLAMMY.
CALCIUM CAREER CLOUDY
CHARGE COTTAGE CHILDREN
CEASAR COMPUTER COMMITTEE
CALCULATOR. DISTANCE DANGLE
GELATIN GUIDE GALLOWS
GAP GAME GOPHER
GRANDMA GREAT. GEESE
GADGET GASOLINE GALLANT
GRAMMER HAMPER HARMONICA
HINGE HIPPOPOTAMUS HURRICANE
HAMMER. IMPORTANT LUMPY
LANGUAGE LUSCIOUS LENGTH
LETTUCE LITTLE LEAD
MEASURE MOTHER MAGNETIC
MADONNA MAGGOT MANIPULATE
MATERIAL METHOD. MASTER
MICROCHIP NEGOTIATE NEUROLOGIST
NUMERICAL OCCUPATION OUTRAGEOUS
POPULATION PEARLY PARTICIPATION
PASSENGER RECIPROCAL RECOLLECT
RUMMAGE SUPPER SCHOOL
STAGGER STRANGE TARGET
TRAIL YEAST YEARN
YESTERDAY YOURSELF YOUNG

THE CHILD MANIPULATED HIS MOTHER INTO LETTING HIM HAVE MACADAMIA NUTS.

SID DROPPED THE LETTUCE INTO GELATIN AND HAD TO PUT ON GALOSHES TO GET IT OUT.

HE WENT TO GUITAR CAMP AND MET A SILLY HIPPOPOTAMUS.

MOTHER HAS CREDIT CARDS AND SHE GOES SHOPPING AND USES THEM AND SHE PAYS A MONTH LATER.

MICHAEL HAS THE GAIT OF A GALLOPING SHEEP.

THE CHIMP NAMED MAY ATE AN ADDITIONAL HELPING OF LEMONS AND LIMES.

PAUL THE GOPHER DANGLES METEORS OUT THE SIDE OF THE COMMUTER TRAIN.

Group 4

F di•di•*dah*•dit
W di•*dah*•*dah*
B *dah*•di•di•dit
J di•*dah*•*dah*•*dah*
, *dah*•*dah*•di•di•*dah*•*dah*
/ *dah*•di•di•*dah*•dit

FND,I	WAO/P	BHCTU
JLRGF	,ECNI	/RSBU
EFIY.	OW,.M	CBDLT
IJEUY	O,.SD	H/NAP
RTFA.	SWULR	ABNCD
NJE.M	CM,DO	I/RY.
IOFLG	WADHS	MD.GS
AJNRP	O,HCY	DO/ER
EFTRS	CUWY.	DBIAH
OJNAY	L,MGD	R/A.I
YMFNI	RWSHL	O.BET
JHCAM	H,DGT	OR/HP
ISRPF	HNAEW	Y.DRB
OTMPJ	CNR,M	DGO/S
FOH.P	ANWRM	LMGEI
BSD.Y	UJRNG	S,RTU

ET/UY	IPGFN	U.MLW
SBRAH	YDCJS	OAN,U
LGDS/	F.IHY	UWOMT
AIBDS	SCUJ.	O,NRC
NAO/Y	UMFSR	TDWHG
INBY.	SAEJR	DMG,I
EI/MN	AF.YO	I,SUD
ALGEBRA	BENJAMIN	BUBBLE
BALLOON	BLUEBERRY	BECAUSE
BRAIN,	BRANCH	BULLETIN
BUTTERFLY	BIRDNEST	FRIGHT
FABULOUS	FANTASTIC/	FACTORY
FAIRY	FAMILY	FAREWELL
FATHER.	FARTHER	FEATURE
FEUD	FIDDLE	FIFTY
FINE,	FIGURE	FLOWER
FLOOD	FEAR	FLASH
JUNCTION	JELLYFISH.	JIGSAW
JOURNAL	JUICE	JUMP
JUNGLE	JUSTICE/	JUSTIFY
LABYRINTH	SANDWICH	TOWARDS
UNDERWEAR	WHERE	WILD
WINDOW,	WAIST	WEATHER
WHETHER	WHALE.	WALLET
WAR	WASP	WHEAT
WHEEL	WITCH/	WHITE
WIFE	WING	WIRE
WORM.	WAGON	WONDER
WALRUS	WHO	WATER
WIFFLE	YAK	YELLOW

LITTLE MISS MUFFET SAT ON HER TUFFET, EATING HER CURDS AND WHEY.

THE FRIGHTENED FATHER/MOTHER WHIP GREEN SMURFS UNTIL THEY TURN BLUE.

A PURPLE CREATURE WITH FANGS/WINGS/FEET WASHED THE WINDOW WITH AN ALCOHOL SWAB.

YESTERDAY, MY FAMILY DOCTOR, WHO IS A WITCH, GASSED UP HIS BROOM AND SAID FAREWELL.

OUR LUNCH MENU CONSISTS OF PLUMS, CORN CHIPS, GARLIC SAUCE, LOBSTER, BROCCOLI AND COTTAGE CHEESE.

THIS HALLOWEEN YOU HAVE A CHOICE OF A GORILLA/ HAMSTER/FLY OR JELLYFISH COSTUME.

THE BATTLE OF BIRDNEST FIELD WAS FOUGHT AT THE JUNCTION/INTERSECTION OF WALRUS ROAD AND MAPLE STREET.

Group 5

K *dah*•di•*dah*
Q *dah*•*dah*•di•*dah*
X *dah*•di•di•*dah*
V di•di•di•*dah*
Z *dah*•*dah*•di•dit
? di•di•*dah*•*dah*•di•dit

RKIDY	NQH.S	OXDAU
TV,FW	BZ/YM	U?NFL
NOK,P	GMWQ/	TDSXY
BVF,R	OSZ.M	PGN?E
AICKU	OPFQ/	FGNXR
ADEVB	HMZI.	WOR?C
UKH/L	WEQIY	MG,XN
FEHCV	ZLBON	D?N/F
BOY.K	U/QHY	FXU.P
MVDWF	SZDIY	MN,P?
UATKG	CQNTE	WLMX.
OCUV/	DBLZU	HEF?N
OHK,S	FWQPM	NX.UD
FVERB	Z.MAB	NEU?Y
CKFAT	OQ/WS	N/XPB
VONS,	IDZ/P	C?UAR
WKB,H	QDEN.	MFWCX
SRTFV	HILPZ	SNMF?
BFKS.	NTUQ/	OX.FU
L/VDY	RBZ,D	T?CFP
DGLKB	NFEQ.	TDBX,

SNFUV	OARZF	ET?BH
TK/GW	QBLNE	MXDTA
AZTEC	EXCITE.	EXERCISE
EXTINCT	EXTRACT	EXAGGERATE
EXASPERATE	EXCELLENT	EXCHANGE
EXECUTIVE,	EXPENSIVE	EXCUSE
FREQUENCY	HERTZ/	KANGAROO
KATHY	KENTUCKY?	KETTLE
KICK	KINGDOM	KIDNEY
MASQUERADE.	QUICK	QUAKER
QUADRUPLE	QUALITY	QUANTITY
QUARTZ	QUEST/	QUILT
QUOTE?	SQUID	SEXTANT
TEXAS	VACCINE	VACUUM
VAIN	VAPORIZE,	VALLEY
VARIABLE	VANDAL	VARIETY
VEGETABLE	VELVET	VERTICAL
VIBRATE.	VICE VERSA	VICTIM
VIOLATION	VIRTUE,	VALVE
VIRGINIA	VITAMIN	VISION/
VOICE	VOLCANO	VOWEL
VULTURE	XRAY?	XEROX
XYLOPHONE	ZOOM	ZEBRA
ZEPPELIN	ZIG ZAG	ZODIAC
ZOMBIE	ZUCCHINI	ZWIEBACK

THE AZTEC PEOPLE BELIEVED IF YOU ATE ZUCCHINI, YOU WOULD TURN INTO A ZOMBIE.

WHEN RADIO FREQUENCIES BECOME HIGH, DO YOUR EYES STING AND YOUR EARS VIBRATE?

THE ERUPTING VOLCANO CAUSED EXTENSIVE VEGETABLE LOSS AND VARIOUS ANIMALS ZIG ZAGGED THROUGH THE FOREST TO AVOID ITS EXASPERATING NOISE.

DINOSAURS ARE EXTINCT BECAUSE THEY ATE ZWIEBACK/ZUCCHINI/SQUASH AND OTHER SUCH VEGETABLES.

WE PUT ZINC OXIDE ON THE WINDOWS BECAUSE THEY FREEZE IN ZERO DEGREE WEATHER, WITHOUT QUESTION.

ZACHERY IS RETIRED FROM THE FIGURE SALON AND NOW HE QUILTS BALLOONS EXACTLY AS VIOLET, ZELDA AND KAREN TAUGHT HIM.

KAREN QUILTED A VERY EXPENSIVE KING SIZE BLANKET, AFTER WHICH SHE XRAYED THE ZONE NEAR THE ZIPPER.

Group 6

1 di•*dah•dah•dah•dah*
2 di•di•*dah•dah•dah*
3 di•di•di•*dah•dah*
4 di•di•di•di•*dah*
5 di•di•di•di•dit
6 *dah*•di•di•di•dit
7 *dah•dah*•di•di•dit
8 *dah•dah•dah*•di•dit
9 *dah•dah•dah•dah*•dit
Ø *dah•dah•dah•dah•dah*

A2JQ7	F8XOR	9YLUH
4,3NM	T/S5U	Z14.H
G6C7K	DP13F	R9V?T
W,5NJ	B6D8I	Q.47U
BAOJM	6R/3C	78LKG
ND1B3	S9E9C	D2H?U
3JIP.	R,6WO	EV8WN
TJ54F	A.6XY	Z4TOP
8OFNS	U,6CA	H9K6Y
P1M3H	VQ2/L	?4E2A
U.QED	R71DJ	QL3W,
5P1G3	T6SOV	A/8WX
GOPX5	1ØDZ.	UM3LN
Z2NOQ	74KMS	R,Ø62
I?S3N	B54HU	TS6LN
W.3YE	SN4KE	X/19L

C7,EN	U8F9R	VN.62
143UT	?TAOB	1C2Q3
B/98K	1,4MO	J5R.9
U?D2L	B76NV	WØ1PB
J35BT	OP7L1	NT.3X
YA/8R	XSØW3	2FJ?B
DOT74	3MRXZ	SB,6P

THE SQUIRREL POPULATION IN THE COUNTRY WAS 4,867,340 IN 1986, 4,654,221 IN 1987 AND 4,320,875 in 1988 BECAUSE OF 54,700 HUNTERS.

AN ESTIMATED 14,736,823,906 RAINDROPS FALL DURING AN AVERAGE STORM/CLOUDBURST IN A 15 SQUARE MILE ZONE, DONT THEY?

IF 115,482 COWS CAN GIVE 724,932 QUARTS OF MILK, HOW MANY 20 GALLON CANS/CONTAINERS WOULD BE NEEDED TO HOLD THAT AMOUNT?

AN AVERAGE 613 SQUARE MILE AREA OF AMAZON RAIN FOREST CONTAINS 1,617,543,219 TREES WITH A TOTAL OF 346,927,004,562 LEAVES.

THE AVERAGE SWAMP CONTAINS 2,865 DIFFERENT SPECIES OF INSECTS, TOTALLING 64,801,952 INSECTS IN ALL, CAPABLE OF BITING 4,923 PEOPLE 612 TIMES.

86 DIFFERENT AIRLINES FLY 415,823 PASSENGERS 13,621,044 PASSENGER MILES, DONT THEY?

HOW COULD A FLOCK OF 1,276,947 BATS EAT 46,339,571 INSECTS WITHIN A 2 HOUR AND 36 MINUTE PERIOD?

Group 7

In actual practice, the only punctuation marks used to any great extent are the period, comma, question mark and slash bar. This next group of characters contains punctuation marks used less often. Learn them to the best of your ability. But don't be discouraged if you find them difficult to recall due to their infrequent usage.

Quotation (“)	di•*dah*•di•di•*dah*•dit
Semicolon (;)	*dah*•di•*dah*•di•*dah*•dit
Hyphen (-)	*dah*•di•di•di•di•*dah*
Colon (:)	*dah*•*dah*•*dah*•di•di•dit

A;LT.	6”KZE	O-NWX
:P8SU	13-HO	V:7NE
D-LAM	3T:OS	5K;RE
TN4”H	A2I-J	OCPU:
E;O1T	V.02?	M-K/Q
V,BLY	SN”F7	4-X.M
G.KXY	4:FW?	L6;EB
N,3KA	Ø-QPI	T/L89
RD2:E	Ø7/NJ	S;FOE
LR?DB	I.SUV	D-1XA
K,LPN	UM:XE	J47.T
1F”GH	LO:BK	V/621
C?VUC	G”FCA	;3XUR
O.QIE	S3/F2	M-FIC
7:HSD	N,2EF	O/R-B
G.NW2	Ø12XL	A:GYZ
F-N2T	H;KEM	Ø4”SL
P6-UN	O?BGI	V.3-R
T”HEQ	:KMWX	6;JNO
13-R.	TV,4L	R/O5J
A:L,3	M-O7U	Q?1EX
V.F-O	R”IEG	XZ;TB
G/2FA	H:RVX	M.LRY

“DIDNT 14 CATS/DOGS/MICE RUN DOWN THE ROAD?” SHE SAID.

1-27/43 IS THE SERIAL NUMBER; AB/24-6 IS THE MODEL NUMBER.

SHE SAID THE FOLLOWING: “ONCE 23 PEOPLE LIVED HERE; THEN 22 PEOPLE LIVED HERE; NOW 18 PEOPLE LIVE HERE.”

ZANEY THINGS HAPPEN, SOMETIMES WITHIN MILES/FEET/INCHES OF EACH OTHER — SO THEY SAY.

"QUILTING 67 QUILTS/BLANKETS TAKES TIME; QUILTING 92 QUILTS/BLANKETS TAKES MORE TIME" SHE SAID.

15-20 IS THE RANGE OF NUMBERS AS FOLLOWS: 15, 16, 17, 18, 19, 20.

HE SAID THE FOLLOWING: "67, 68, 69 ARE NUMBERS THAT ARE CONSECUTIVE; 71, 83, 90 ARE NUMBERS THAT ARE NOT."

Group 8

This last group of characters are special symbols used to indicate various conditions during telegraphy conversations. Where several characters are shown with a line drawn across the top, they are to be sent as a *single* character. For example, $\overline{AS}$ is transmitted as di•*dah*•di•di•dit, *not* di•*dah* di•di•dit.

Error	di•di•di•di•di•di•di•dit
Message received OK (R)	di•*dah*•dit
Invitation to transmit (K) (to any station)	*dah*•di•*dah*
Invitation to transmit ($\overline{KN}$) (only to station in contact with)	*dah*•di•*dah*•*dah*•dit
Wait ($\overline{AS}$)	di•*dah*•di•di•dit
Break ($\overline{BT}$)	*dah*•di•di•di•*dah*
End of message ($\overline{AR}$	di•*dah*•di•*dah*•dit
End of contact ($\overline{SK}$)	di•di•di•*dah*•di•*dah*
Understood ($\overline{SN}$)	di•di•di•*dah*•dit
Attention ($\overline{KA}$)	*dah*•di•*dah*•di•*dah*

Code Practice Methods

There are a number of ways to practice Morse code, each of which has advantages and disadvantages. This section reviews various methods of practice and shows you how to make use of available training material.

Random Code Generators

This device automatically generates Morse code characters. Advances in digital electronics enable it to incorporate some amazing features. Most units allow selection of the type of characters produced, such as all letters, or letters and numbers. Speed of sending is variable over a wide range, to suit individual needs. Some devices permit sending with different lengths of time between characters (ie, characters sent at high speed with long spaces between). Others allow the operator to program in automatic speed increases. For example, a unit may send at 5 WPM for the first five minutes, 7 WPM for the next five minutes, and so on.

Such devices have a number of advantages. First, they produce perfect code. Also, speed is variable to any desired setting. Finally, it is impossible to memorize transmissions because they are randomly generated by electronic circuitry. Their only disadvantage may be price. They are more expensive than some other methods of practice.

Code Cassettes and Records

Tape cassettes and records are available by mail from a number of different suppliers. Some of these contain random code. Others provide practice in on-the-air telegraphy conversations.

The advantages of using this type of material are its low cost and that it provides perfectly sent (machine-generated) Morse characters. The primary disadvantages are that the speed cannot be regulated and it is possible the listener may memorize the practice material after frequent usage. Also, since the speed cannot be changed, the user is forced to purchase additional tapes as his or her speed increases. If one intends to attain higher code speeds (let's say in the 20-WPM range) then purchase of a different type of practice device, such as a random code generator, may be advisable.

Listening to Another's Sending

The main advantage in this means of training is that a sender can tailor practice to the learner's needs. He can send characters with which the listener is having difficulty, and he can transmit at any speed. This method has a drawback if the sender does not have a good "fist." No one wants to listen to poor sending, and doing so is detrimental to someone trying to learn techniques of good telegraphy. Matters are helped if the sender uses an "electronic keyer." This device is discussed in a later chapter.

One suggestion for the newcomer is to locate an Amateur Radio Club in your area. These clubs often provide free-of-charge code prac-

tice for beginners. Sometimes the names and addresses of such clubs can be found in the "Recreation Directory" provided in many newspapers. A more direct approach is contacting ARRL HQ for help in locating a club in your area. There are some 2000 Amateur Radio Clubs affiliated with the ARRL throughout the US.

Listening on the Air

By listening to on-the-air transmissions, one becomes accustomed to factors involved in monitoring live transmissions. Such things as atmospheric irregularities, interference from other stations and poor sending become apparent. Also, if one already owns a receiver, it makes sense to use it for code practice. To do so costs nothing.

There are many sources of Morse code on the airwaves. Two portions of the radio spectrum that contain an abundance of this type of transmission are the Maritime Services frequencies and the CW portions of the Amateur Radio bands. Both are listed here. Also, see the Reference Section for code practice schedules from ARRL HQ station W1AW.

CW-Only Portion of High Frequency Amateur Bands

3.500 MHz-3.750 MHz (80 meters)
7.000 MHz-7.150 MHz (40 meters)
10.100 MHz-10.150 MHz (30 meters)

CW-Only Portion of High Frequency Amateur Bands (continued)

14.000 MHz-14.150 MHz (20 meters)
18.068 MHz-18.110 MHz (17 meters)
21.000 MHz-21.200 MHz (15 meters)
24.890 MHz-24.930 MHz (12 meters)
28.000 MHz-28.300 MHz (10 meters)

Note: There are no official subbands on 160 meters (1.800-2.000 MHz), but by convention, CW operation occurs in the low end, generally below 1.840 MHz.

Maritime Services Bands

2.050 MHz-2.100 MHz	8.350 MHz-8.700 MHz
4.180 MHz-4.350 MHz	12.300 MHz-13.100 MHz
6.250 MHz-6.525 MHz	16.600 MHz-17.250 MHz

You should not rely solely on any single type of transmission for code practice. Many telegraphy users tend to carry on rather short conversations and some make extensive use of abbreviations. There is quite a difference, for example, in successful copy of each of the following sentences.

SHIPMENT BETWEEN PORTS WILL BEGIN AT 1100 HOURS. PLEASE STAND BY FOR INSTRUCTIONS.

TNX VY MUCH OM. UR RST 599 HR IN EASTON. VY FB QSO. 73 FER NW ES GD LUCK.

Computer Code-Teaching Aids

Rapidly coming to the forefront in code-learning methods is an assortment of computer-code teaching aids. Some of these are designed to provide receiving practice only, but others include training in sending as well.

Morse Tutor, for the IBM PC XT, AT and compatibles teaches all code characters in 11 lessons, using a "flash card" technique for each character. You can set up each lesson to teach just the characters in that lesson, a random character drill using only the characters just introduced or a random-word drill using all of the characters taught through that lesson. Characters can be displayed as they are sent or at the end of the lesson.

The final lesson is a random QSO generator based on a huge pool of information that is contained on the disk. Two stations make a contact with several exchanges of information during each QSO—just like the real thing. The contacts are similar to those used on code exams. Names and callsigns of the stations match throughout the contact, and the lesson can be interrupted by hitting any key. Then, you can start where you left off, or quit at any time.

Morse Tutor is easy to calibrate for different computer clock speeds. Code speeds and character spacing are selected separately, both in WPM, so you can copy regular code or use the Farnsworth method. The program remembers your choice for these variables as well as lesson duration, tone frequency and display mode. Morse Tutor is user friendly, and has easy-to-understand menu-driven functions. Excellent error-trapping and accuracy in the code speed being sent make this software even more attractive.

Morse Tutor is available on 5¼" diskettes at many ham radio dealers or directly from ARRL headquarters for $20.00 plus $3.00 postage and handling ($4.00 for UPS). Morse Tutor is also available on 3½" diskettes for $22.00 plus $4.00 postage and handling ($5.00 for UPS).

Aids for Morse Code Reception

Several items are available that will make on-the-air code copy more efficient and enjoyable. First, a pair of headphones will provide isolation from disturbing background noise. This is especially important when listening to very weak signals. Headphones are a necessity when copying code with a typewriter because of the noise produced by that machine.

Next, an audio filter is a must for any serious code enthusiast. This is one of the greatest devices ever invented for Morse code reception. It is capable of selecting code signals of different audio pitch. This means that under crowded band conditions a single signal can be selected from a multitude of others by its unique tone. Some receivers have built-in filters. If yours does not—buy one! A 500-Hz filter is usually suitable for most CW work, although narrower filters are available.

Autek Research audio filter

Sending

It is important to be able to send good code. No one wants to be called a "lid" (poor sender), so you will want to do well for your personal satisfaction and pride. Good code also makes it easier for a listener to understand your message. You can verify this by listening to various signals on the air. A properly transmitted message is much easier to copy even at a much greater speed than a poorly sent one.

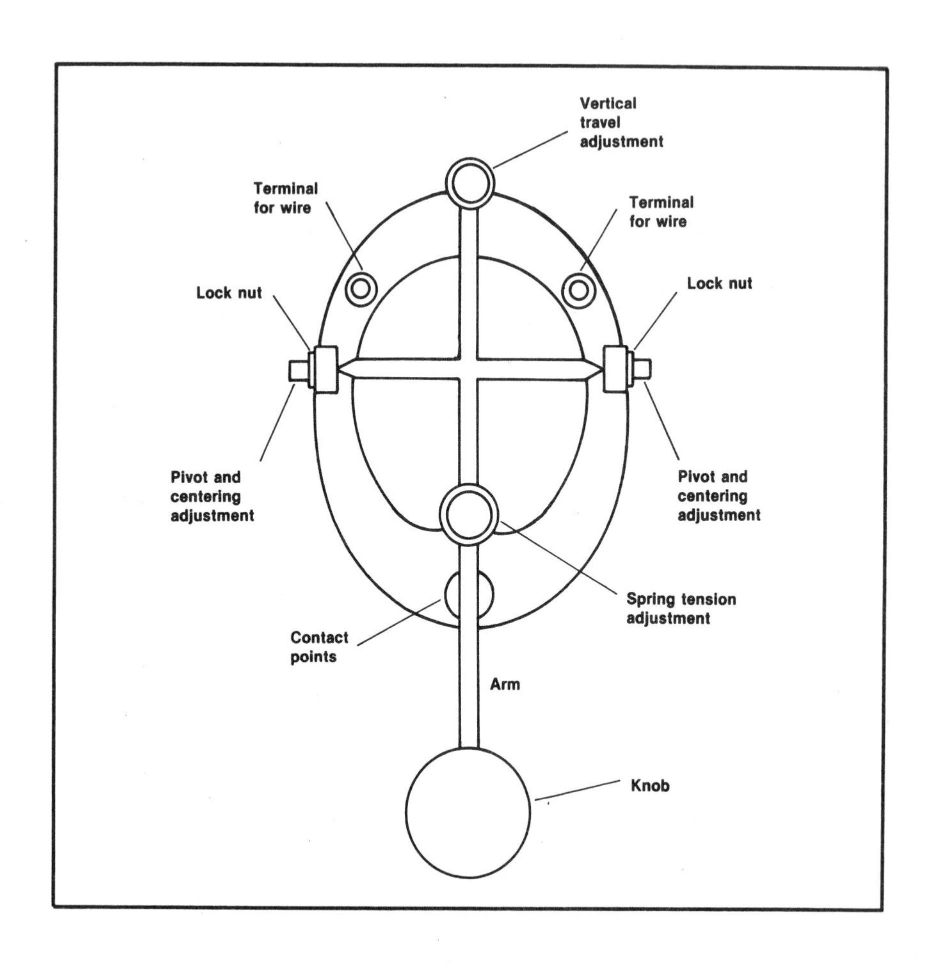

Despite its importance, you should not attempt to send code until you have reached a certain level of competence in receiving. You do not have to be able to receive at high speeds, but you should have a perfect recollection of all characters, to the point where you can state each symbol immediately after your instructor sounds it out. This ability is necessary because when transmitting you cannot afford the time to think of how a character is formed—it must be almost second nature to you. Once you have reached that stage, you are ready to devote your mental efforts to those tasks essential to good sending.

Begin by correctly adjusting your telegraph key. You will not be able to send well without doing this.
Proceed as follows.

1) Adjust both *pivot and centering adjustments* so the *contact points* are aligned one directly above the other. While doing this, be

careful of two things. Do not adjust the pivot and centering adjustments so tightly that the arm cannot move freely up and down, nor so loosely that it exhibits sideplay. Once properly adjusted, tighten both lock nuts to maintain alignment.

2) Adjust the *vertical travel adjustment* so there is 1/16'' vertical travel when measured at the *knob*. Tighten the associated lock nut.

3) Adjust the *spring tension adjustment* to your individual taste. This adjustment regulates the amount of pressure required to depress the key. It should be loose enough that sending can be accomplished with relative ease and without fatigue, but tight enough that the key is not so easily depressed that sending becomes sloppy. Tighten the corresponding lock nut.

These are initial settings. After using the key for a while, you may want to reset the vertical travel adjustment and spring tension adjustment more to your liking. Although it is important these adjustments feel right to you, they can be the cause of sending difficulties. Too large a gap and/or too much tension can cause choppy sending. Too small a gap and/or too little tension can cause characters or character elements to be slurred together.

You will need some type of signal source to sound out the dits and dahs as you send. The most desirable form of device is the code practice oscillator. A number of companies manufacture this piece of equipment. It is a simple audio oscillator to which a telegraph key is connected. Controls are provided to vary tone and volume.

You can also build an oscillator yourself from easily obtainable parts, or a signal source can be constructed from the simple hookup of a buzzer and two batteries. Schematics for both arrangements are provided here.

It is very difficult to break poor sending habits once they start, so be sure to cultivate your sending habits correctly. Begin by mounting your key to a table of standard 30-inch height. It should be affixed about 18 inches from the table edge so your elbow is able to rest on the table when you grip the knob. You may screw it to the table or adhere it with some type of sticky substance such as beeswax. It must be firmly attached so it does not move as you send.

Sit erect, facing the table. Place your elbow on the table and your hand on the key. Your wrist should be off the table, raised slightly by the muscles of your forearm. Grasp the knob as illustrated, with thumb on the left edge, forefinger toward the front and middle finger

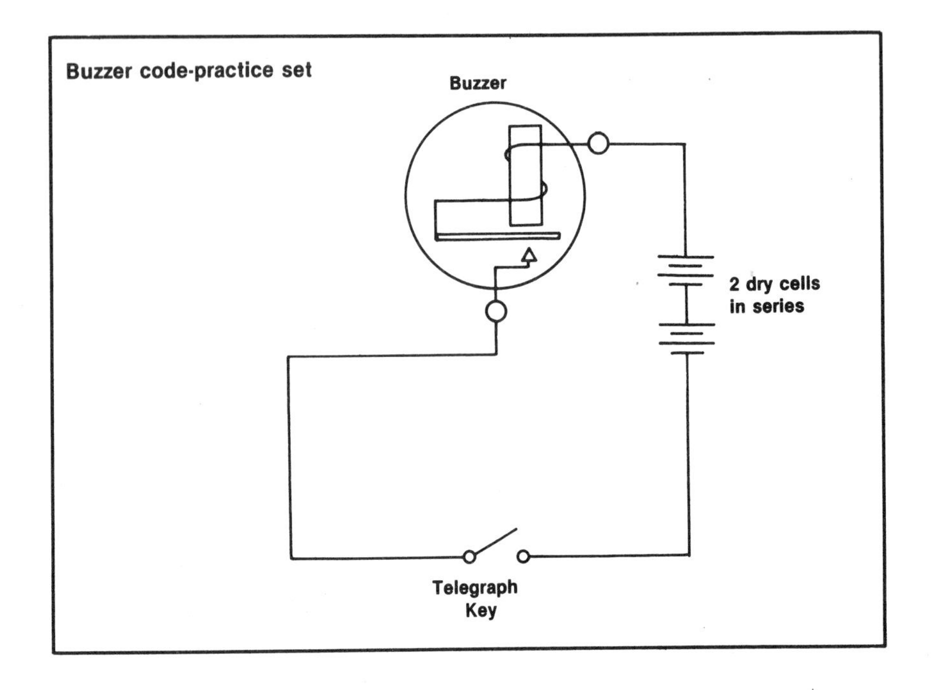

to the right. This is a starting position. After using the key for a while, you may find a different position is more suitable. If so, change to what you feel is a more comfortable grip. The hand, wrist and arm should always be relaxed.

It is easier to send code than to receive. As a result, some beginners make the mistake of starting their sending practice at higher speeds than they should. In so doing, they skip over the basics of proper sending and often become poor senders. Don't let this happen to you. Start your sending program as you did your receiving exercises, with a string of 10 dits.

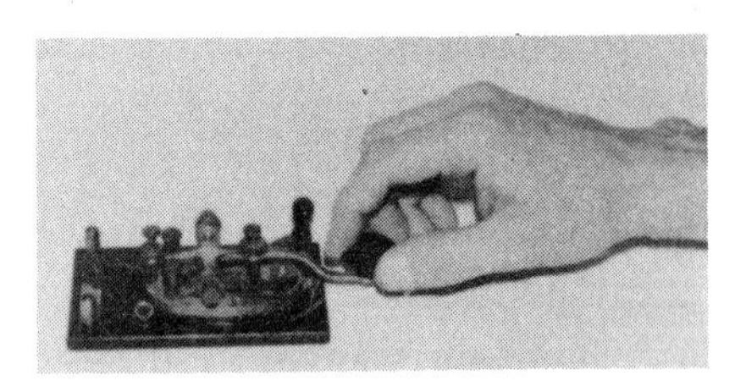

Illustration showing proper grasp of key

di•di•di•di•di•di•di•di•di•dit

Timing is the most important factor in sending. Pay attention to the length of the dits and the space between them. Try it again, four or five more times. Make sure the dits are evenly spaced. Then try a string of dahs.

dah•dah•dah•dah•dah•dah•dah•dah•dah•dah

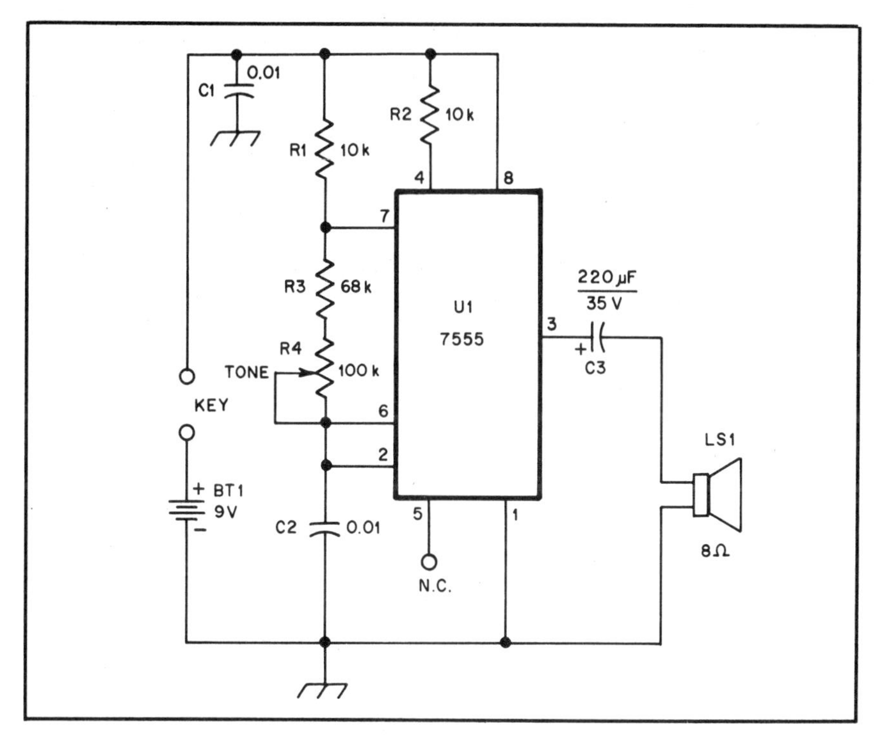

Schematic diagram of code-practice oscillator. Radio Shack part numbers are given in parentheses.

BT1—9 V battery (23-553)
C1, C2—0.01 µF capacitor (272-131)
C3—220 µF, 35 V electrolytic capacitor (272-1029)
LS1—2 inch loudspeaker, 8 Ω (40-245)
R1, R2—10 kΩ resistor, ¼ watt (271-1335)
R3—68 kΩ resistor, ¼ watt
R4—100 kΩ potentiometer (271-220)
U1—555 IC timer (276-1723)

Each dah should be three times as long as a dit, but the spaces between dahs should be the same as those between dits.

Element	*Duration in No. of Dits*
Dit	1
Dah	3
Space between elements	1
Space between characters	3
Space between words	7

Send the string of dahs four or five more times. Get the feel of the key. If the key, or your hand position on the key, feels uncomfortable, make any necessary adjustments. Now try a string of dits and dahs interspersed.

di•*dah*•di•*dah*•di•*dah*•di•*dah*•di•*dah*

Keep the rhythm smooth. This string should sound like *one* long character, not individual As.

Continue these exercises until you feel perfectly comfortable with the key and you feel your rhythm is correct. If available, a metronome can be used to great advantage in helping you attain proper element lengths and spacing.

Another excellent way to attain proper rhythm is to tape record your practice sessions and play them back, listening for any sending faults. If you do this in the beginning, it can help you stop bad sending practices before they become habits. Now try this string of ten numbers.

1 2 3 4 5 6 7 8 9 Ø

Sending practice is best done in groups, similar to receiving practice. Start with the first group shown here and continue to succeeding groups as you feel you have mastered each set of characters. As you practice, strive for good sending, not speed.

Group 1

EISH5 I5HSE H5SIE
H5SIE S5IEH HIS5S
SH5E5 ES5HI IHS5E
H5SE5 H5IES 5IEHS
5E5HS E5HSI SE5SH
E5SIE

Group 2

TMOØT OØOTM TMTØO
OMTOM ØMOTT OTMØM
OMTØT MOTØO OMTØT
TMOØM MTØMO OTMØT
MOØTT ØMØTM OTMØØ
MTTØØ

Group 3

LRWJ1 RW1AP J1PA1
LJPRL PWJRL JWAL1

RWALP	WRALA	1PWRA
RAJWP	W1JPA	JWR1P
JWP1A	P1AWW	ARW1L
WL1RJ		

Group 4

3VUF2	VUV2F	3V4FU
2V2FU	4VFU3	43UV3
VU43F	VFU43	U3VF2
VF2UF	F323U	43FV2
234UF	4VUF2	VUF43
U34FU		

Group 5

NDB6X	BX9D6	9XN86
689ND	96DBN	8NBD8
6NBD8	BX9D6	XBDN8
96DNB	DBNX6	DX98B
69XDN	X6B89	NXBD8
6BX89		

Group 6

YQZ7K	GZQG7	7Z7QC
CKCQG	GQY7Z	YCQCZ
CYQGK	GKYC7	CQGKC
Q7YKZ	KZ7GQ	ZGQ7Z
7ZQGY	ZCQGY	YGQ77
YCY7Q		

Upon completion of all exercises, you may practice sending material from any text. Always maintain your speed at a point at which you feel you have complete control. Strive for good code, and increases in speed will come naturally.

Abbreviations

In order to expedite transmission, many telegraphers make use of abbreviations. Listed here are the more common ones.

AA	All after
AB	All before
ABT	About
ADR	Address
AGN	Again
ANT	Antenna
BCNU	Be seeing you
BK	Break
BKR	Breaker
BN	Been; between
BTR	Better
B4	Before
C	Yes
CFM	Confirm
CK	Check
CKT	Circuit
CL	Call; closing down station
CLD	Called
CLG	Calling
CLBK	Callbook
CLR	Clear
CONDX	Conditions
CNT	Cannot
CQ	A general call to any station
CU	See you
CUD	Could
CUL	See you later
CUM	Come
DA	Day
DE	From
DIFF	Difference
DLVD	Delivered
DN	Down
DUNNO	Don't know
DX	Distance
EL(S)	Antenna element(s)
ES	And
FB	Fine business (indicates approval)
FER	For
FM	From
FREQ	Frequency
GA	Good afternoon
GB	Good bye
GD	Good
GE	Good evening
GESS	Guess
GG	Going
GM	Good morning
GN	Good night
GND	Ground
GV	Give
GVG	Giving
HI	Telegraphic laughter
HPE	Hope
HQ	Headquarters
HR	Here; hear
HV	Have
HW	How
INFO	Information
LID	Poor telegraph operator
LNG	Long
LTR	Later; letter
LV	Leave
LVG	Leaving
MILL	Typewriter
MSG	Message
N	No
NCS	Net control station
ND	Nothing doing
NIL	Nothing; I have nothing for you
NR	Number
NW	Now
OB	Old boy
OM	Old man
OP	Operator
OPR	Operator
OT	Old timer
PKG	Package; parking
PSE	Please
PT	Point
PWR	Power
PX	Press

R	Roger (received OK)
RC	Rag Chew
RCD	Received
RCVR	Receiver
RE	Concerning
REF	Refer to
RIG	Station equipment
RPT	Repeat
RST	Readability, strength, tone
RX	Receiver
SASE	Self-addressed, stamped envelope
SED	Said
SEZ	Says
SGD	Signed
SIG	Signal; signature
SKED	Schedule
SRI	Sorry
SS	Sweepstakes (an on-the-air contest)
STN	Station
SUM	Some
SVC	Service; prefix to service message
SW	Switch; shortwave
TFC	Traffic
TMW	Tomorrow
TKS	Thanks
TNX	Thanks
TR	Transmit
T/R	Transmit/Receive
TRIX	Tricks
TT	That
TTS	That is
TU	Thank you
TX	Transmit; transmitter
TXT	Text
U	You
UR	Your
URS	Yours
VFB	Very fine business (indicates approval)
VY	Very
W	Watt; watts
WA	Word after
WB	Word before
WD	Word
WDS	Words
WID	With
WKD	Worked
WKG	Working
WL	Well; will
WRD	Word
WUD	Would
WX	Weather
XMTR	Transmitter
XTAL	Crystal
XYL	Wife
YL	Young lady
YR	Year
73	Best regards
88	Love and kisses

Q Signals

Q signals are special, three-letter abbreviations beginning with the letter Q that have recognized meanings for radio operators the world over. This makes possible limited conversations between operators of different nations who do not understand each other's native language. Q signals were originally devised for ship-to-shore communications but are now employed for many other types of CW contacts. A Q signal can have dual meaning. If it is not followed by a question mark, it is an affirmative statement. If it is followed by a question mark, it asks a question. Some common Q signals used in Amateur Radio operating are listed below:

	Affirmation	*Interrogation*
QRG	Your exact frequency is . . .	What is my exact frequency?
QRH	Your frequency varies.	Does my frequency vary?
QRI	The tone of your signal is: 1. Good 2. Variable 3. Bad	How is the tone of my signal?
QRJ	I cannot receive you. Your signals are too weak.	Are you receiving me badly? Are my signals weak?

	Affirmation	*Interrogation*
QRK	The intelligibility of your signal is: 1. Bad 2. Poor 3. Fair 4. Good 5. Excellent	What is the intelligibility of my signal?
QRL	I am busy.	Are you busy?
QRM	I am being interfered with by other stations.	Are you being interfered with by other stations?
QRN	I am troubled by static.	Are you troubled by static?
QRO	Increase your power.	Shall I increase my power?
QRP	Decrease your power.	Shall I decrease my power?
QRQ	Send faster, at...WPM.	Shall I send faster?
QRS	Send slower, at...WPM.	Shall I send slower?
QRT	Stop sending.	Shall I stop sending?
QRU	I have nothing for you.	Do you have anything for me?
QRV	I am ready.	Are you ready?
QRW	Tell...that I am calling him on...(frequency).	Shall I tell...that you are calling him on ...(frequency)?
QRX	I will call you again at ...hours (on... (frequency)).	When will you call me again?
QRZ	You are being called by...	Who is calling me?
QSA	The strength of your signal is: 1. Scarcely perceptible 2. Weak 3. Fairly good 4. Good 5. Very good	What is the strength of my signal?
QSB	Your signal is fading.	Is my signal fading?
QSD	Your keying is defective.	Is my keying defective?
QSK	I can hear you between my signals.	Can you hear me between your signals?
QSL	I acknowledge receipt.	Do you acknowledge receipt?
QSO	I can communicate with ...direct (or by relay through...).	Can you communicate with ...direct or by relay?
QSW	I am going to send on this frequency. Or, I am going to send on ...(frequency).	Will you send on this frequency? Or, Will you send on... (frequency)?

	Affirmation	*Interrogation*
QSX	I am listening to. . . on. . .(frequency).	Will you listen to. . . on. . .(frequency)?
QSY	Change to transmission on another frequency. Or, Change to transmission on. . .(frequency).	Shall I change to transmission on another frequency? Or, Shall I change to transmission on. . .(frequency)?
QSZ	Send each word or group twice. Or, Send each word or group . . .times.	Shall I send each word or group twice? Or, Shall I send each word or group. . .times?
QTH	My location is. . .	What is your location?
QTR	The correct time is. . . hours.	What is the correct time?
QTV	Stand by on. . .(frequency) from. . .to. . .hours.	Shall I stand by for you on. . .(frequency) from. . .to. . .hours?
QTX	I will keep my station open for further communication with you until further notice. Or, I will keep my station open for further communication with you until . . .hours.	Will you keep your station open for further communication with me until further notice? Or, Will you keep your station open for further communication with me until. . .hours?

Being familiar with these Q signals will enhance the operating skill of the average amateur. But there are other Q signals. For example, if you ever become involved with "net" operation, the following list will be helpful. A net is a group of radio operators operating on the same frequency for a common purpose.

Special Q Signals for CW Net Use

QNA†	Answer in prearranged order.
QNB†	Act as relay *B*etween. . .and. . .
QNC	All net stations *C*opy.
QND†	Net is *D*irected.
QNE†	Entire net please stand by.
QNF	Net is *F*ree.
QNG	Take over as net control station.
QNH	Your net frequency is *H*igh.
QNI	Report *I*nto net or I report into net.
QNJ	Can you copy. . .?
QNK†	Transmit message(s) (for. . .) to. . .
QNL	Your net frequency is *L*ow.
QNM†	You are QRMing the net. Please stand by.

QNN	Who is net control station? *N*et control station is...
QNO	Station is leaving the net.
QNP	Unable to copy you (or...).
QNQ†	Move frequency to...and wait for...to finish handling traffic. Then send him traffic for...
QNR†	Answer...and *R*eceive) traffic.
QNS†	Following *S*tations in net...
QNT	I request permission to leave the net for...minutes.
QNU†	The net has traffic for YOU: await instructions.
QNV†	Establish contact with...on this frequency. If successful, move to...kHz, and send him traffic for...
QNW	How do I route messages for...?
QNX†	You are excused from net or I request to be excused.
QNY†	Please shift to another frequency (or to...kHz) to clear traffic with...
QNZ	Zero beat your signal with mine.

†for use only by the Net Control Station.

Many other Q signals exist. While not usually used by amateurs, they are listed here to provide a reference source in case you should encounter unfamiliar ones. Incidentally, don't let these lists frighten you. The average amateur makes use of a dozen or fewer Q signals. You'll easily pick up those commonly in use after listening to on-the-air QSOs.

	Affirmation	*Interrogation*
QAV	I am calling...	Are you calling me?
QCM	There seems to be a defect in your transmission.	Is there a defect in my transmission?
QDH	The present interference is being caused by...	What is causing the present interference?
QIF	...(station) is using ...(frequency).	What frequency is... (station) using?
QRA	The name of my station is...	What is the name of your station?
QRB	I am about...from your station.	How far are you from my station?
QRD	I am going to...and I am from...	Where are you going and where are you from?
QRR	I am ready for automatic operation. Send at... WPM.	Are you ready for automatic operation?
QRY	Your turn is number...	What is my turn?
QSG	Send...telegrams now.	Shall I send...telegrams now?
QSJ	The charge to be collected per word to...including my internal telegraph charge is...	What is the charge to be collected per word to... including your internal telegraph charge?

	Affirmation	*Interrogation*
QSM	Repeat the last telegram that you sent.	Shall I repeat the last telegram I sent?
QSN	I heard you on. . .(frequency). Or, I heard. . .on. . .(frequency).	Did you hear me on. . . (frequency)? Or, Did you hear. . .on. . . (frequency)?
QSP	I will relay to. . .free of charge	Can you relay to. . .free of charge?
QSQ	I have a doctor on board. Or, I have. . .on board.	Do you have a doctor on board? Or, Do you have. . .on board?
QSU	Send or reply on this frequency. Or, Send or reply on. . . (frequency).	Shall I send or reply on this frequency? Or, Shall I send or reply on . . .(frequency)?
QSV	Send a series of Vs on this frequency. Or, Send a series of Vs on . . .(frequency).	Shall I send a series of Vs on this frequency? Or, Shall I send a series of Vs on. . .(frequency)?
QTA	Cancel telegram number . . .as if it had not been sent.	Shall I cancel telegram number. . .as if it had not been sent?
QTB	I do not agree with your count of words. I will repeat the first letter or digit of each word or group.	Do you agree with my word count?
QTC	I have. . .telegrams for you. Or, I have. . .telegrams for. . .	How many telegrams do you have to send?
QTE	Your TRUE bearing from me is. . .degrees (at. . .hours). Or, Your TRUE bearing from . . .was. . .degrees (at . . .hours). Or, The TRUE bearing of. . . was. . .degrees (at. . . hours).	What is my TRUE bearing from you? Or, What is my TRUE bearing from. . .? Or, What is the TRUE bearing of. . .from. . .(call sign)?
QTG	I am going to send two dashes of ten seconds each followed by my call sign (repeated. . .times) (on . . .(frequency)). Or,	Will you send two dashes of ten seconds each followed by your call sign (repeated. . .times) (on. . .(frequency))? Or,

	Affirmation	*Interrogation*
	I have requested...to send two dashes of ten seconds each followed by his call sign (repeated ...times) (on...(frequency).	Will you ask...to send two dashes of ten seconds each followed by his call sign (repeated...times) (on...(frequency))?
QTI	My TRUE track is...	What is your TRUE track?
QTJ	My speed is...	What is your speed?
QTL	My TRUE heading is... degrees.	What is your TRUE heading?
QTN	I departed from...at ...hours.	When did you depart from ...?
QTO	I have left port. Or, I am airborne.	Have you left port? Or, Are you airborne?
QTP	I am going to enter port. Or, I am going to land.	Are you going to enter port? Or, Are you going to land?
QTQ	I am going to communicate with you by means of the International Code of Signals.	Can you communicate with me by means of the International Code of Signals?
QTS	I will send my call sign for...minutes now (on ...(frequency)) so that my frequency may be measured. Or, I will send my call sign for...minutes at... hours (on...(frequency)) so that my frequency may be measured.	Will you send your call sign for...minutes now (on...(frequency)) so that your frequency may be measured? Or, Will you send your call sign for...minutes at ...hours (on...(frequency)) so that your frequency may be measured?
QTU	My station is open from ...to...hours.	During what hours is your station open?
QUA	Here is news of...(call sign).	Have you any news of... (call sign)?
QUB	Here is the information you requested.	Can you give me information concerning visibility, height of clouds, direction of wind and velocity of wind at...(place of observation)?
QUC	The number of the last message I received from you is... Or, The number of the last message I received from ...(call sign) is...	What is the number of the last message you received from me? Or, What is the number of the last message you received from...(call sign)?

	Affirmation	*Interrogation*
QUD	I have received the urgency signal sent by . . . at . . . hours.	Have you received the urgency signal sent by . . . ?
QUF	I have received the distress signal sent by . . . at . . . hours.	Have you received the distress signal sent by . . . at . . . hours?
QUG	I am forced to land at . . . (position or place).	Will you be forced to land?
QUH	The barometric pressure at sea level is . . .	What is the barometric pressure at sea level?

CHAPTER FIVE

High Speed Operation

What is "high speed" code? Generally, speeds of greater than 20 WPM are considered high speed. Of course, such a definition is subject to individual interpretation. Some may consider 20 WPM to be super-fast; others may see it as just loafing. Perhaps the best way to get speed into perspective is to look at some of the lowest-acceptable, average and world-record figures for code reception.

Five WPM seems to fit well as a lowest-acceptable figure, since this is the minimum allowable speed for passing any radio license exam (Novice Amateur Radio license).

The average speed of Amateur Radio operators, the largest group of code users in the world, is perhaps 15 WPM. This is an estimate, for speeds of individual operators will vary from 3 to 65 WPM, while some may be unable to receive code at all, after years of disuse.

The world's record for code reception was set on July 2, 1939 by Ted R. McElroy. His amazing record of 75.2 WPM has stood ever since. McElroy earned the title of Official Champion Radio Operator at the Asheville Code Tournament in Asheville, North Carolina.

Levels of speed referred to are usually those of receiving rather than sending. With recent developments in code sending devices, the upper limit of man's ability to send code is actually the upper limits of man's ability to type. There is one sending record, however, that is of interest. According to the *Guinness Book of World Records*, the record for sending code by use of a straight key is 35 WPM. It was set on November 9, 1942 by Harry A. Turner, a member of the US Army Signal Corps, who accomplished the feat at Camp Crowder, Missouri. This record is significant because it demonstrates an upper limit for straight-key work and points out the need for other sending instruments capable of higher speeds.

As with many other records, reports of some code proficiency records vary as to names, dates and places. J. B. Milgram, in the October-November-December 1982 issue of *Dots and Dashes*, the official journal of the Morse Telegraph Club, reports the following records of code reception and transmission by straight key.

- 1855—Jimmy Leonard, 15 years old, received code at a rate of 55 WPM at the Paris Exposition. The code was sent by Joseph Fisher, who used a straight key.
- 1866—Billy Kettles, 18 years old, transmitted code at the rate of 40½ WPM at the first official telegraph competition held by the American Telegraph Company.
- 1890—B. R. Pollack sent code at 52 WPM at the Hardman Hall Fast Sending Tournament in New York.

These records appear to be superior to the one referenced in the *Guinness* book possibly because years ago a word was considered in many contests to be comprised of four characters, instead of five, as it is today.

Improving Your Receiving Speed

Whether you have aspirations of setting a new world's record or just improving your receiving skills, there are certain things you can do to attain your goals more quickly.

One of the most important transformations in the ability to copy occurs when one stops listening to individual letters and begins to hear "word sounds." When this happens an operator becomes oblivious, for example, to the individual letters I, N and G in the word KEYING, and instead hears an ING ending only. Similar situations occur with other parts of words and complete words. Portions of words like ED, EY, ONS, ION and CON begin to sound like entities in themselves. The same thing happens with many smaller words such as THE, AND, ON, OF, WHEN, WAS, INTO, BEEN, FOR, HAVE, CAN, YOU,

HAS, AFTER, HE, IN, WHAT, WOULD, VERY, and many more.

The ability to recognize words and sections of words is the secret of all good, high-speed operators. When it begins to happen to you, you will suddenly appreciate the Morse code for what it really is—another language. This skill will develop automatically as you advance, but there are ways to speed up the process. It is absolutely essential that this process occur, for the average operator begins to have difficulty distinguishing individual characters at a speed of 25 to 28 WPM.

One way this skill can be cultivated is to listen to code practice of characters sent at a high rate of speed with long pauses between each, as discussed in chapter 4. This forces you to begin hearing word sounds instead of letter sounds. This does not mean you will hear *all* words or sections of words as separate entities, but your vocabulary of sounds will gradually become larger and larger. As you work toward your speed goal, each word sound you recognize will become a relaxing break in your copying endeavor. If you listen to the sentence "I ran to the door to see what would happen," and you have developed a certain skill with word sounds, fully one-half of the sentence will require little effort to copy. The "TO," "THE," "WHAT," "WOULD" and "PEN" sounds will appear familiar to you, and you will be able to write down these portions with little thought effort.

An important part of the development of this skill is the ability to *spell*. Since you will no longer be listening for separate letters, the spelling of words will be up to you. For example, if you hear the *sound* of the word WOULD, but do not detect the component letters, it would be your responsibility to spell it W-O-U-L-D when you write it down. WOULD is easy to spell, of course, but it serves to illustrate the point.

The best way to improve your receiving skills is to listen to code that is transmitted at a rate of speed slightly faster than you are able to copy. Doing this challenges you to exert the effort necessary to bring your speed up to the speed of code transmission. You should start at a speed at which you are able to copy 85 to 90 percent of what is sent. Stay at that level until your copy improves to almost 100 percent, then increase the speed of transmission again. This is somewhat easier said than done; it is not always possible to find code transmissions at the exact speeds which you require. One way to accomplish this is to work with a partner. There is another method, however, which is probably the best means of practice available for either high-speed or low-speed code exercise. This method makes use of a tape recorder.

Many manufacturers sell battery-operated, portable cassette recorders. Some are battery-only types; others have facilities to operate from household current. As long as a recorder is battery operated, the motor driving the tape unit must be a low-voltage, dc motor. With little effort, a speed control can be inserted in the wiring to such a motor, and the result will be a variable-speed tape recorder!

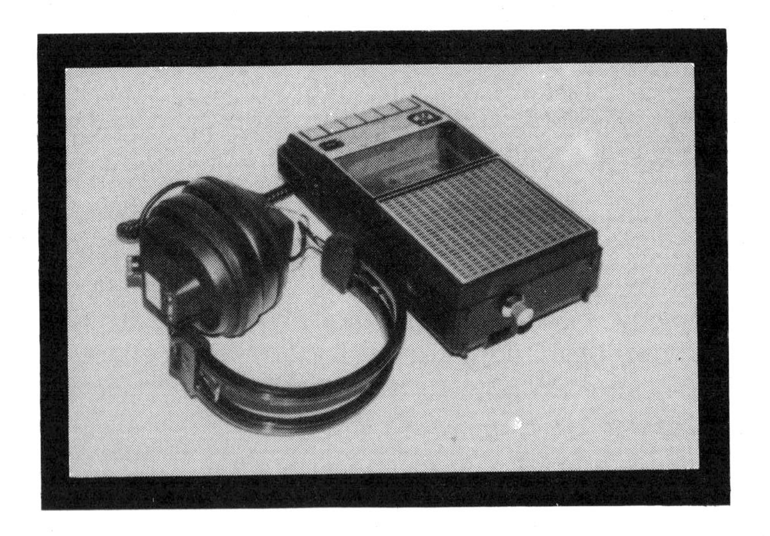

Pictured here is the author's unit, which was modified by insertion of a single 35-ohm variable resistor in series with the motor wiring. Other arrangements are possible. For example, if you find a low-ohmage, variable resistor difficult to locate, you can use a multiple-position switch in conjunction with fixed resistors. And you need not insert the speed control in the recorder itself if you are reluctant to drill a hole in the unit. Wiring can be arranged so the control is mounted externally, in a separate enclosure. The exact hookup and location of parts is up to you. The overall objective is to reduce the voltage applied to the motor by inserting a resistance between the motor and its voltage source. Be careful not to disturb any other circuitry.

If the device is operated from batteries only, there is no danger from shock. If you operate your recorder from the ac line, then proper wiring and construction practices must be observed to be certain no safety hazards exist.

Once a recorder is so modified, it can be used to advantage in a number of ways. You can purchase prerecorded tapes at high speeds and slow them down to your level, gradually increasing speeds of practice runs as you improve. Or, you can record your own tapes of live, on-the-air CW runs, or for that matter, of your own sending. Depending upon the speed of transmission, these runs might be taped with the recorder at one-half speed, for example, so there is plenty of leeway for increases in playback speed as your receiving abilities improve.

Some individuals have the ability to copy CW "in their heads." That is, they are able to remember most, if not all, of a CW transmission without putting it down on paper. Such talent is to be admired, but there are times when all of us, including those with this skill, must record what is received. Perhaps the exact content of a transmission has to be relayed to another person. Or, for some reason a record of a particular message may be needed for future use. You may intercept an urgent distress call with pertinent information as to times, dates and places. Or, you may be involved in a code-copying contest. Suffice it to say it is important that everyone has the ability to document a CW message.

Most operators copy messages by hand either by printing or writing. However, the upper limit of an individual's ability to copy in this manner is about 40 WPM. Therefore, one is forced to use a typewriter, or "mill," if he or she desires to receive code above this speed. Also, some may wish to use a typewriter so their messages are neat and easy to read.

The first requirement for doing this is the ability to type. While it is true one can hunt-and-peck through a message, such an approach severely limits progress. If you do not already know how to type correctly, learn how to do so. Self-study books are available and so are evening courses at many schools.

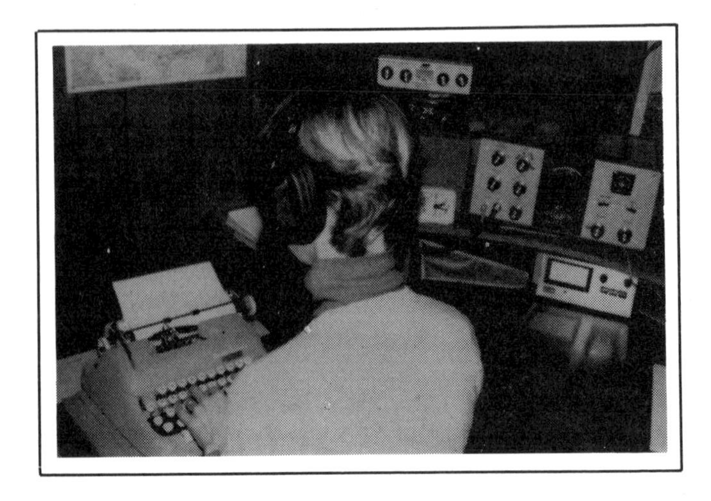

The biggest problem that arises when using a typewriter is noise. As a result, it is imperative that you use headphones. Find a pair with large ear cushions. These serve to keep out noise and simultaneously provide wearing comfort. Set your typewriter on a separate table from your receiver. This prevents vibrations from disturbing the receiver and possibly causing shifts in frequency. Place a sponge-rubber pad under the machine. This can be purchased at any typewriter store. In addition to reducing noise, a pad prevents the machine from shifting during use.

When copying by typewriter, you may experience difficulty at first coordinating your hands with your brain. This is normal, but you will find that any awkwardness disappears after several practice sessions. You will soon be at ease using a mill, just like the pros.

Strive for *smooth* copy. Do not type fast for several words, slow down or stop, then speed up again. Get into a natural rhythm. The only way this can be done is by copying slightly behind the text being sent. That is, as you listen to one word, you type the previously sent word. As a matter of fact, good operators copy not one word behind, but three, four or five words behind. At first, attempting to copy in this manner will be very difficult. Doing so requires the ability to dissociate the mind from physical actions. In other words, one must carry out one process consciously while doing another subconsciously.

Such a skill can be acquired. For example, you can try copying code while simultaneously carrying on a conversation with someone. Or, you can write down one list of words while reading another list. As with all your success to this point, practice is the real key.

Check the Reference Section for lists of on-the-air code practice sessions. Speeds are listed so you can pick those runs which most closely suit your needs.

Above all, don't become discouraged. You may hit more plateaus as you progress; this is common. Stick with it. You will eventually overcome each plateau and will soon be receiving at speeds you can boast about. You may even find strange, encouraging things happening to you, such as being able to copy better at higher speeds than lower ones! Some individuals have commented they find this to be the case because at higher speeds word sounds become more obvious. At low speeds it is easy to pick out the T, H and E in the word THE. But it is easier to hear the entire word sound of THE at high speeds. As letters become less discernable, word sounds become more discernible. You should notice this as you approach higher speeds.

As the realm of very high speed code is approached, most of us will, of course, encounter increased difficulty in our attempts to break ever higher speed barriers. The author has tried some unorthodox approaches to overcoming his limitations, and has met with varying degrees of success. Some of these attempts are being passed along with the hope that certain individuals will benefit from them.

To begin, the lower end of 40 meters is an excellent place to find some real pros in the CW game. Many high-speed operators occupy the spectrum between 7000 and 7050 kHz. It has become a popular "hangout" for the CW aficionado. Listening here will not only provide some excellent high-speed practice, but also lessons in proper CW operating techniques. Skills of some of the amateurs found here are truly amazing.

Whether listening to the bottom end of 40 or using some other means of code generation, try to mentally visualize the code as being printed out in front of you on a sheet of paper as you hear it. Some find this to be helpful.

Another trick is to turn up the speaker volume on the device being used for code practice so that it may be heard as you walk around the house. Then, use half your brain to concentrate on the work at hand and the other half to decipher the code (without writing it down). Obviously, this technique can only be used when doing something that does not require your full concentration. And it is best done when family members are not home! Practice such as this not only aids in learning the code but helps to improve one's skill of disassociating the mind from physical activities.

Still another idea, and one that the author has found to be particularly helpful, is to listen to code at a considerably faster rate than you are able to receive. Although perhaps only one-third to one-half of the words may be understood, this form of learning forces one to hear entire words and word sounds instead of individual letters. The process can be compared to the Farnsworth method in which the individual more easily recognizes letter sounds due to the high rate of speed at which the letters are sent. In addition, there is the added advantage in that when the speed is reduced to the point where the operator is able to copy 100 percent, the rate of transmission will suddenly appear slow to the listener. Some might describe this benefit as being "psychological," but even if this is so, it *is* a benefit and it *does* help.

Also, the **Code Practice Schedules** section of the Appendix provides information on some very high speed (up to 60 WPM) on-the-air code practice transmitted by W1NJM from Newington, Connecticut.

If your ambition is to someday set a new world's speed record, there are high-speed telegraphy competitions for this purpose, although tournaments of this type are not common. Perhaps the best known is the International Amateur Radio Union's contest.

The International Amateur Radio Union (IARU) is a federation of national Amateur Radio societies representing the interests of two-way amateur communication. The IARU is divided into three regions. Secretariats for Region 1 are in England, for Region 2 in Canada, and for Region 3 in Japan.

Region 1 sponsors a high-speed competition every so often. It first took place in Moscow in 1983 and its second in Hannover, Germany in 1989. While plans for future events are uncertain, current recommendations are that world championships shall take place not more frequently than once every two years, and that regional championships shall be held every other year.

There were two basic categories for the 1989 world championships, a tournament of individual competitors and a team competition consisting of three to six members per team. Competitions were of two types, a High-Speed Competition and an Open-Class Competition.

The High-Speed Competition consisted of the following tests.

1. Reception of letter messages.
2. Reception of figure messages.

3. Transmission of letter messages.

4. Transmission of figure messages.

The Open-Class Competition consisted of the tests below.

1. Reception with copying of mixed text messages (letters, figures and punctuation marks) for three minutes and reception with copying of plain English text messages for two minutes.

2. Reception with memory copying of three plain English text messages.

3. Transmission of mixed text messages (letters, figures and punctuation marks) for three minutes and transmission of plain English text messages for two minutes.

Perhaps someday *you* will be competing in one of the above championships!

Mechanical Sending Devices

The straight key was used for years as the standard device for transmitting code. But as needs arose to relay more and more messages, faster methods of keying were devised.

Although many different types of telegraph keys have been invented, only a few have enjoyed wide-spread use. The "bug" is one of these, and it will be discussed below. The "sideswipe" falls in line shortly after the straight key and bug as one of the more common mechanical keys. This key consists of a paddle mounted on an arm which is pivoted so the arm can move sideways. Contact points are mounted on both the right side and left so that a key-down condition exists when the paddle is pushed to either side. An operator moves the paddle back and forth with his thumb on one side and his index and second finger on the other side, in whatever fashion he wishes, to tap out a message. This mechanism, although never finding wide acceptance, does have an advantage over a straight key because sending is somewhat faster and takes less effort.

A bug is the most popular type of high speed, mechanical telegraph key. Just how the bug derived its name is not certain. One story holds that an operator was using one for landline telegraphy during the 1800s, not realizing his very rapidly transmitted message interrupted another's conversation. To the operator who was being interfered with, the rapid sending (especially the dits) sounded like a bug, and he irately commented "Get that bug off the line!"

Regardless of how it received its name, the bug was popular for years, and is still used by some operators today. This device consists of a paddle mounted to an arm which is pivoted so it can move sideways, very similar to a sideswipe. The difference between the bug and sideswipe is that the bug makes dits automatically. This is accomplished through use of a leaf-spring mounted at the middle of the bug's arm. When the paddle is moved to the right, and held to the right, the leaf-spring vibrates back and forth, alternately making and breaking an electrical contact which is mounted farther down the arm. Dashes must be made manually by moving the paddle to the left.

Vibroplex bug (photo courtesy Vibroplex Company, Inc.)

To send well, it is important that the bug is adjusted properly. Proceed as follows.

1) Clean all contacts with a contact cleaner. Use a burnishing tool only if the contacts are pitted.

2) Loosen up the dit contact adjustment, left side arm stop adjustment and right side arm stop adjustment screws.

3) Screw in the right side arm stop adjustment screw to the point where it *just barely* moves the rear portion of the arm, but not so far that it moves the front portion of the arm away from the arm shock damper. When this is done, the arm should be perfectly straight. To test it, move the paddle to the right very slowly. As it is moved, the entire arm should remain straight, with the arm leaf-spring not flexing at all. This adjustment is rather critical. Its objective is to prevent any arm bounce when the arm is released after pushing the paddle to the right to make dits. Test it out.

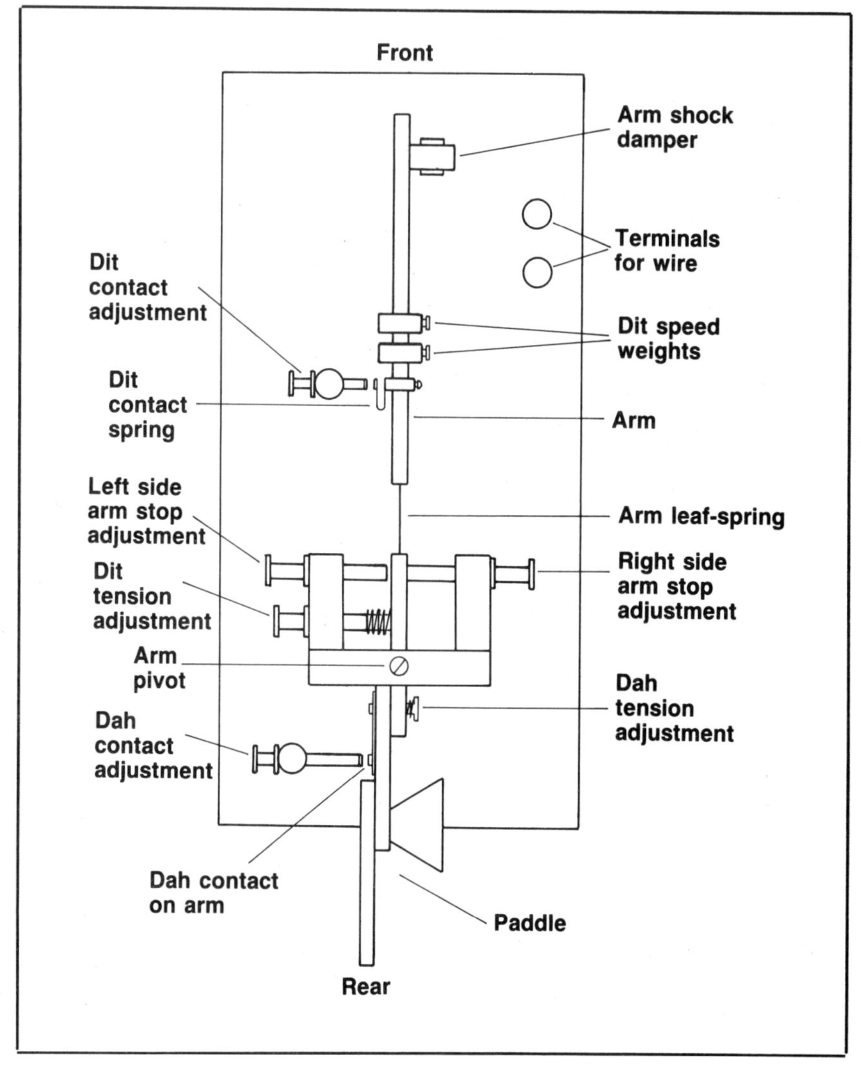

4) Very slowly push the paddle to the right until the arm is forced against the left side arm stop adjustment screw. Adjust the left side arm stop adjustment so there is approximately 1/8 inch between the front portion of the arm and the arm shock damper.

5) Move the dit speed weights toward the front of the arm, close to—but not touching—the arm shock damper.

6) Push the paddle to the right as far as it will go, making a string of dits, and hold it there until it stops vibrating completely. Adjust the dit contact adjustment so it just touches the dit contact spring. Make sure points on the dit contact adjustment screw and dit contact spring align together perfectly. Next, place an ohmmeter across the two terminals for wire. Move the paddle quickly to the right, making a series of dits. Fine-tune the dit contact adjustment so the average deflection of the ohmmeter is mid-scale for the first five to ten dits only. Trying to read the meter for dits past the tenth one will give erroneous readings because the dits become increasingly heavy as the paddle is held in position. (Note: This method of fine tuning the dit contact adjustment by use of an ohmmeter appeared in February 1968 *QST* in the article "Semi-Automatic Key Adjustment" by Brian Murphy, VE2AGQ).

7) Adjust the dit tension adjustment to your personal preference.

8) Adjust the dah contact adjustment so there is approximately 1/64-inch distance between the contact point on the arm and the contact point on the dah contact adjustment screw. Make sure the set of points aligns exactly. The 1/64-inch travel is an average figure, and can be altered to suit your liking.

9) Adjust the dah tension adjustment to your personal taste.

10) Adjust the dit speed of the bug to the speed at which you wish to send by sliding the dit speed weights toward the front or rear accordingly.

If you experience difficulty while sending, review the following list for the possible cause.

Choppy sending

1) Too short or too long a travel in the dit or dah paddle movement.

2) Improper tension in dit or dah paddle movement.

Choppy dits

1) Dirty contacts.

2) Dit paddle movement too short.
3) Dit contact adjustment out too far.

Heavy dits

1) Dit contact adjustment in too close.

Dits die out too quickly

1) Dit paddle movement too short.

After, and *only* after, your bug is properly adjusted should you attempt to send with this instrument. Position the bug in a location similar to that of your straight key. Many operators locate these two keys directly next to each other on their operating table. Make sure it is firmly attached to the table top in some manner. Beeswax placed under the feet works well and will not damage the operating surface.

Warm up to the key by sending a series of dits, then a series of dahs. Try sending some strings of numbers next; they are excellent practice for bug use. Start off *slowly*. The biggest mistake new operators make is to begin racing away with their new "plaything." There are several reasons for proceeding at a moderate pace. First, you are less likely to pay attention to good sending practices if you start off too fast. Also, you should not let your sending speed outpace your receiving speed. If for no other reason, you are sure to be embarrassed when another operator returns your call at the same speed you are sending, and you find yourself unable to copy his transmission!

Operate relaxed. You should send with a gentle, *rolling motion* of the wrist. Most operators grip the paddle with their thumb on the left side and their index and second fingers on the right. There is no single, correct way to hold the paddle or to send; these are suggestions for the beginner.

Now, more than ever, you must pay attention to proper element and character spacing. There is nothing wrong with personalized code. Every operator's sending has its own, unique characteristics. This is fine, as long as the personalization is not overdone to the point where poor code is the result.

Electronic Receiving and Sending Devices

Major advances have been made in keying and code-reading

devices since the advent of the transistor and integrated circuit. Some of these instruments have truly amazing capabilities.

Electronic Keyers

While the bug remains the mainstay of a select group of operators, virtually all serious CW operators have switched to the electronic keyer. This device is similar to the bug in that a sideways paddle movement is used to make dits and dahs. However, it has many advantages over its mechanical counterpart.

Palomar Engineers electronic keyer (photo courtesy Palomar Engineers)

The paddle of an electronic keyer is pushed to the right to make dits and to the left to make dahs. The dits are produced automatically, as with the bug, except that they are formed perfectly because they are electronically generated. The dahs are also produced automatically, which is not the case with the bug. As a result of the automatic generation of character elements and perfect element spacing, the electronic keyer enables an operator to transmit excellent code.

There are two basic types of electronic keyers. One uses a built-in paddle, the other an external paddle. Each has its advantages. The type with built-in paddle allows a somewhat neater station appearance, while an external paddle gives the operator more freedom in positioning his keying device and takes up less space on the operating table.

Vibroplex keyer paddle (photo courtesy Vibroplex Company, Inc.)

Almost all electronic keyers manufactured today are *iambic,* or squeeze keyers. This unit is identical to those just described with one exception. It uses a dual-handle paddle which has the ability to make a consecutive series of di•*dah* or *dah*•dit combinations with a single squeeze of the two handles. For example, if you squeeze the handles together, hitting the left handle just a split-second before the right one, a di•*dah*•di•*dah*•di•*dah* string will be continuously emitted. If you squeeze in the same manner, but hit the right handle before the left one, a continuous *dah*•di•*dah*•di•*dah*•dit string will be the result. This feature enables the operator to make many characters with less hand movement.

A C can be sent with a single squeeze of the paddle, by hitting the *dah* lever first. The result will be *dah*•di•*dah*•dit. Then, both paddles must be released just after the last dit has begun or the string of alternating *dahs* and dits will continue. The period can also be transmitted with a single squeeze, by hitting the dit lever first, resulting in di•*dah*•di•*dah*•di•*dah*.

Transmitting a character with other than continuously alternating dits and *dahs* requires a little more skill. Using the F as an example, the dit lever is hit first and held in place. After the start of the second dit is heard, the *dah* lever is tapped and then quickly released. This inserts a *dah* after the two dits. Upon releasing the *dah* lever, the dits will continue. Therefore, the dit lever must be released after the start of the final dit is heard. The result is di•di•*dah*•dit. The L is transmitted in a similar manner, by holding down the dit lever continuously while tapping the *dah* lever at the appropriate point in time. Q is transmitted by holding down the *dah* lever continuously and tapping the dit lever when needed.

If you are considering the purchase of an electronic keyer, you should be aware of the options available on different models. All but the simplest of these units have an output jack for keying a transmitter, built-in audio oscillator for monitoring your keying, and volume, tone and speed controls. Additional options include:

1) Separate outputs for keying different types of transmitters (usually solid state or non solid state transmitters).

2) Headphone jack.

3) Jack for external key.

4) Connector to route the audio output from a receiver into the keyer.

5) Internal memory to record messages.

6) "Weight" control.

The last item deserves special attention. A weight control adjusts the ratio between the dit and dah lengths and the length of the space between dits and dahs. You may find this to be a strange control, for ratios between element and space lengths are fixed in the Morse code and they have been drilled into you since the code was first presented.

However, unusual things can happen at high speeds. Above 25 WPM, some operators perceive that the code sounds somewhat choppy with standard ratios, and they prefer other settings. The exact ratio used is one of individual preference. If you purchase a keyer with this option, try various settings. You may find your code sounds smoother with weight settings different from the norm.

Some operators find it difficult to adjust to an electronic keyer, especially those who previously used a bug. To make certain you encounter as little difficulty as possible, test out the feel of the paddle before you buy. There are wide differences in the feel of keyer paddles. Choosing the one that's right for you can make a big difference in the quality of code you send.

Although all keyers have a speed control, few are calibrated in words-per-minute. There is one trick that can be used by those who are interested in knowing exactly what speed their keyer is set to. This hint was originally presented in May 1980 *QST*, in an article by Jim Pitts, K4EY.

Obtain a pocket calculator of the type that displays a running total each time the = key is depressed. To see if yours is of this type, key in 1 + 1 and then hit =. The display will show 2. Now hit = again. If the display goes to 3, you have the right type of calculator. Locate the two wires going to the = key. Cut the wires and solder an extension wire to each. Attach these extension wires across your keyer output. Now key in 1 + 1. Then, send a string of dits for exactly one minute. Then note the total displayed. Divide the total by 25 and you can determine your keyer speed setting in words-per-minute. If you experience difficulty in obtaining a total, try reversing the leads to the keyer.

CW Keyboards

As sophisticated as these keyers seem, there are devices even more advanced. The CW keyboard is one of these. Indeed, it is hard to imagine a device that could make the sender's job easier.

The keyboard of this machine is identical to that of a typewriter, with a few exceptions. There are usually a number of keys that perform special functions. For example, one key may send the symbol for end-of-QSO ($\overline{\text{SK}}$), and another the symbol for wait ($\overline{\text{AS}}$).

The two most important items in selection of a keyboard are the feel of the keys and the amount of "buffer" area provided. Just as there are poorly made typewriters and computer terminals, there are poorly made CW keyboards. These are usually immediately evident from the feel of the keyboard, and such units are to be avoided. Choose a keyboard that feels right. Often, price is an indication of quality.

Most keyboards (all of the better ones) are buffered. This means they have an electronic storage area that temporarily stores the information you type. This information is then released at the rate at which you have the keyer speed set. As an example, suppose you type at 50 WPM, but you have the keyer speed set to transmit at 25 WPM. The information you type will be released at the lower rate. This is done to "smooth out" the information you key and produce perfect code. The CW keyboard does, in fact, produce absolutely perfect code. It is a joy to listen to.

Since the information is temporarily stored, it is important to have a sufficiently large buffer, especially if you are a fast typist. Some keyboards have no buffer area, some have a small area (perhaps suitable for holding 15 characters), others have a large area (capable of holding 500 characters or more). Buffer storage costs money. Buy as much as you can afford.

Many keyboards have the capacity to perform a long list of functions. Controlling these various functions is accomplished in one of two ways. Some units have switches and knobs mounted on the keyboard face. Others control almost all operations by the use of special combinations of keys. For example, depressing a key labelled CNTRL while simultaneously pushing the 1 key, may perform one function. Depressing the CNTRL key and 2 key at the same time may perform another function. Listed here are the options and capabilities of some keyboards.

1) Ability to key different types of transmitters.

2) Buffer status indicators (tell when buffer area is half full and completely full, for example).

3) Ability to store and recall messages that are used frequently.

4) Capacity to produce test messages, such as "The quick brown fox jumped over the lazy dog's back."

5) Tape cassette interface.

6) Weight control.

DGM Electronics CW keyboard (photo courtesy DGM Electronics)

7) Control of intercharacter spacing.

8) Built-in, random code generator.

9) Backspace key. (Erases incorrectly typed information before it is transmitted).

10) Ability to automatically fill in any pause in transmission with the break symbol ($\overline{BT}$).

Automatic Code Readers

Automatic Morse code readers are available that can detect code and spell out messages in a constantly moving, lighted display. These devices attach to the speaker or headphone output of a receiver. Despite their amazing capabilities, however, no machine has yet been devised that is capable of copying very weak or poorly sent Morse as effectively as the human ear.

Home Computer Interfaces

Several companies manufacture an interface that enables an operator to transmit and receive code through a personal computer connected to a transceiver. The Advanced Electronic Applications (AEA) PK-232 Multimode Data Controller is an example of such a unit. The PK-232 has the capacity not only to transmit and receive Morse, but also RTTY, ASCII, AMTOR (Amateur Teleprinting Over Radio), Packet (a system of digital communication in which information is broken into short bursts, or "packets") and Facsimile (a means of sending drawings, charts, maps and graphs), and it can copy the Japanese and Russian versions of the Morse Code! It is a truly versatile and amazing piece of equipment.

To accomplish all this, the PK-232 uses several programs supplied on floppy disk. Programs are available from AEA for the IBM PC and compatibles and for the Commodore. Other types of computers, such as the Apple, may be accommodated through the use of software sold by the computer manufacturer. Almost every type of computer can be made to work with this unit.

Connections are made to the computer, transceiver and power supply and the necessary programs are loaded. Then, various commands are fed into the computer keyboard to control operation. During transmit the keyboard is buffered so that keying may be done at higher speeds while the actual transmission is released at lower speeds and is sent as perfectly spaced Morse Code.

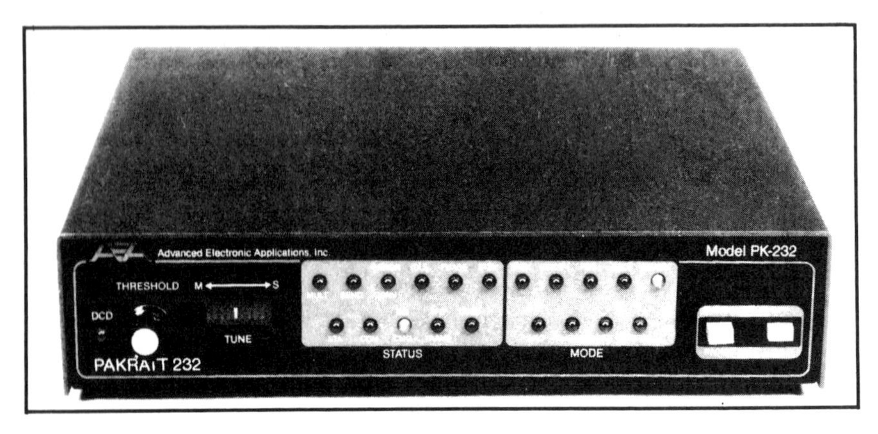

AEA PK-232 Multimode Data Controller

A split screen CRT display is used so that transmitted data and received data can be shown simultaneously. During reception the incoming portion of the QSO is stored on disk for future retrieval. It should be noted that this unit, despite the fact that it is computer-controlled, will not receive poorly sent Morse. It is best used to receive machine-sent code or well spaced hand transmissions.

Phillips Code

At some point in your amateur career, you will come across the words "Phillips Code." In the late 1800s, the telegraph was widely used by the news services. Owing to the large volume of information that had to be relayed, telegraphers made extensive use of abbreviations. In 1879, a high-speed operator named Walter P. Phillips published the first edition of the Phillips Code, a comprehensive list of abbreviations, many of which were derived from popular abbreviations in use at the time. The code became a standard for the industry, and through its use, many telegraphers of the day were able to clear 15,000 words per shift! It was to their advantage to handle as much information as possible; it was common to be paid by the word or message. Today this code is rarely used.

An interesting letter written in 1940 by L.R. (Mac) McDonald, W8CW, to Bill Pierpont, later licensed as NØHFF, gives some insight into the Phillips Code and the world's record for code reception held by Ted McElroy.

Detroit, Michigan
August 29, 1940

Wm. G. Pierpont
304 S. Rutan,
Wichita, Kans.

Dear Bill:

Thanks for your inquiry about Phillips code.

It was found many years ago that handling press matter at the speed demanded by newspaper organizations was too much strain on the operators. An operator named Walter Phillips introduced a system of shorthand which could be transmitted over a circuit. The operator on the receiving end spells everything out in full. Phillips is handy for sending naughts, hundred is sent HND, thousand TND, million MYN, trillion TYN, etc. Several words can sometimes be sent by a few letters, such as POTUS for President of the United States.

The publisher is Telegraph and Telephone Age, 66-70 Beaver St., New York City. The book costs $1.50. You can get it from the publisher direct, or I can supply you.

If you have studied the Candler courses there is not much I can tell you about code. Incidentally Walter Candler revised one of the editions of the Phillips Code many years ago and was one of the best Phillips Code operators in the country. Guess you know Walter died last April. His wife is carrying on the course and is doing fine with it. She is a former Western Union operator. I took Candler's course in 1919.

About the speed tests, government count is used, that is five units to the word. Only plain newspaper English is used, everything having clear meaning. No trick stuff. In the Asheville tournament, the speed was practically the same for McElroy and myself. We both copied solid at 75 w.p.m. (press matter prepared by the F.C.C.) but they sent us some stuff at 77 ½ w.p.m. and I didn't get a good start on in. McElroy made something that looked like copy, but pretty ragged looking, so they gave him 75.2. I guess it was. If only first class copy had been counted it would have ended in a tie. McElroy and I have had about the same telegraph experience.

Sincerely,

L.R. McDonald — W8CW
11421 Grandmont,
Detroit, Michigan

Some abbreviations from the Phillips Code are included in here. The complete list is many times longer. You can see by studying this list how some of our currently used abbreviations were derived.

Phillips Code Abbreviations

ABB	Abbreviate	HRD	Heard	POS	Possible
AGR	Agree	HRT	Hurt	PRM	Permanent
ALG	Along	IA	Iowa	QA	Qualify
APP	Appoint	IFM	Inform	QNY	Quantity
APRL	April	INC	Increase	QRL	Quarrel
BAJ	Badge	ITN	Intention	R	Are
BF	Before	IW	It was	RF	Refer
BH	Both	JL	Jail	RLY	Really
BLU	Blue	JP	Japan	RPT	Repeat
BWR	Beware	JUN	June	SB	Subsequent
C	See	KD	Kind	SG	Signify
CFM	Confirm	KG	King	SNT	Sent
CK	Check	KM	Communicate	TBL	Trouble
COL	Colonel	KN	Known	TH	Those
CR	Care	LAK	Lake	TKT	Ticket
DA	Day	LGR	Longer	U	You
DED	Dead	LIB	Liberty	Un	Until
DMH	Diminish	LV	Leave	UR	Your
DRU	Drew	MA	May	VA	Virginia
EA	Each	MD	Made	VU	View
EDU	Educate	MK	Make	VY	Very
EGO	Egotism	MO	Month	WB	Will be
ENR	Enter	N	Not	WD	Would
EV	Ever	ND	Need	WH	Which
FA	Fail	NR	Near	XAC	Exact
FEV	Fever	NTC	Notice	XM	Extreme
FQ	Frequent	OB	Obtain	XPT	Export
FZ	Freeze	OFN	Often	XU	Exclude
GD	Good	OMN	Omission	Y	Year
GNT	Grant	OT	Out	YD	Yield
GVT	Government	PAS	Pays	YOA	Years of age
GW	Grow	PB	Probable	ZA	Sea
HAP	Happy	PD	Paid	ZD	Said
HH	Has had			ZM	Seem

CHAPTER SIX

Proper CW Operating Practices

If there were one most-important piece of advice for the CW operator, it would have to be: *Do a lot of listening!* Listening gives the operator a feel for the band. Who is on the air; what geographic area are the majority of signals coming from; what are band conditions; is there a lot of QRM, QRN or QSB; Are there any particular unused spots on the band that would be a good place to call CQ. Get a feel for the airwaves before taking action.

A contact may be initiated in one of two ways, by answering a CQ or by sending one. It is best to answer someone else's CQ if possible. Why have numerous CQs occupying the airways when the objective of each is the same — to establish a contact with another station.

Whether sending a CQ or answering one, *be brief.* Don't stretch out the procedure unnecessarily. Doing so not only wastes time but reflects poorly on one's operating abilities.

When answering a CQ, send the other station's call several times followed by DE and your call repeated several times. W1AAA W1AAA DE W1BBB W1BBB $\overline{\text{AR}}$. $\overline{\text{AR}}$ signifies end of message. When listening on the air it is not uncommon to hear $\overline{\text{KN}}$ used in place of $\overline{\text{AR}}$, but this is not recommended. $\overline{\text{KN}}$ is used to indicate that only a particular, contacted station is to return your call. Since contact has not truly been made at this point in time, $\overline{\text{AR}}$ is more appropriate.

After contact has been established and a QSO is in progress, $\overline{\text{KN}}$ should be used. Also, each call need be repeated only once from this point on when returning control to the other station. W1AAA DE W1BBB $\overline{\text{KN}}$ is adequate. In fact, FCC rules require that a transmitting station identify itself only once every 10 minutes, so if transmissions are brief, a simple BK (for break) can be used to indicate that it is the other party's turn to transmit.

An example of an intelligently initiated CQ would be as follows: CQ CQ CQ CQ DE W1AAA W1AAA CQ CQ CQ CQ DE W1AAA W1AAA K. Notice that the string of CQ's and the number of times it is repeated are both minimal.

When ending a contact, $\overline{\text{SK}}$ is used to signify end of QSO in the following manner: $\overline{\text{SK}}$ W1AAA DE W1BBB.

Dialogue within the QSO itself will, of course, be comprised of anything that is on the operators' minds at the time. Discussions could range from the latest baseball scores to the naming of a baby or the fact that snow is due tomorrow. But before the conversation reaches this point, it is customary to provide each other with three basic pieces of information, a signal (RST) report, name of the operator and station location (QTH).

A sample QSO, using proper CW procedures and kept short for demonstration purposes, would look like the following.

CQ CQ CQ CQ DE W1AAA W1AAA CQ CQ CQ CQ DE W1AAA W1AAA K

W1AAA W1AAA DE W1BBB W1BBB $\overline{\text{AR}}$

W1BBB DE W1AAA GA OM ES TNX FER THE CALL. VY NICE SIG IN HR TODAY. UR RST 599 599. NAME HR IS JIM JIM ES QTH IS BOSTON MA BOSTON MA. HW CPY? W1BBB DE W1AAA $\overline{\text{KN}}$

(Good afternoon old man and thanks for the call. Very nice signal in here today. Your signal report is 599. Name here is Jim and location is Boston, Massachusetts. How copy?) Note that it is common practice to repeat important information such as signal report, name and location twice.

W1AAA DE W1BBB FB JIM. UR RST 599 599 ALSO. NAME HR IS FRED FRED ES QTH IS PORTLAND ME PORTLAND ME. WX TODAY IS BEAUTIFUL ES PLAN SUM FISHING LATER THIS AFTERNOON. W1AAA DE W1BBB $\overline{\text{KN}}$

(Fine business Jim. Your signal report is 599 also. Name here is Fred and location is Portland, Maine. Weather today is beautiful and plan some fishing later this afternoon.)

W1BBB DE W1AAA R I LOVE TO FISH ALSO BUT GET LITTLE TIME FER IT. WISH I CUD GO WID U HI. SRI BUT MUST RUN, AM WORKING NITE SHIFT ES HVE TO GET READY. TNX VY MUCH FER QSO ES 73. $\overline{\text{SK}}$ W1BBB DE W1AAA CL

(Roger, I love to fish also but get little time for it. Wish I could go with you [laughter]. Sorry but must run, am working night shift and have to get ready. Thanks very much for the contact and best regards.) (I am closing down my station.)

W1AAA DE W1BBB OK JIM, WILL LET U GO. DONT WORK TOO HARD HI. TNX FER CALL ES 73. $\overline{\text{SK}}$ W1AAA DE W1BBB

(OK Jim, will let you go. Don't work too hard [laughter]. Thanks for the call and best regards.)

It should be noted here that although the above is technically correct, many operators eliminate the period after a sentence and insert extra spaces instead. This makes for faster communications, and if done properly is quite understandable. Without using periods, the last message from W1BBB above would look like this:

W1AAA DE W1BBB OK JIM, WILL LET U GO   DONT WORK TOO HARD HI   TNX FER CALL ES 73 $\overline{\text{SK}}$ W1AAA DE W1BBB

Some additional practices will aid in making your CW activity more enjoyable and will show others that you are a savvy operator. For example, keeping a list of Q-signals and CW abbreviations handy can prevent you from being embarrassed by not understanding a portion of another's transmission. Of course, learning the most often used Q-signals and abbreviations is a necessity.

Always send a QRL? (Is this frequency in use?) before sending a CQ. This will prevent you from transmitting on a frequency that may appear to be vacant, but in fact is being used by others. This practice is especially important on certain bands such as 10 meters, where it is common to hear only one side of a transmission because of propagation conditions.

Be willing to slow down (QRS) to a speed that the other operator is comfortable in receiving. And, it is best to call someone who is sending a CQ at approximately a speed which you are able to copy without too much difficulty. To expect an operator sending CQ at 40 WPM

The "Considerate Operator's Frequency Guide"

Some frequencies that are generally recognized for certain modes or certain activities (all frequencies are in MHz):

160 Meters	
1.800-1.830	CW, RTTY and other narrowband modes
1.830-1.840	CW, RTTY and other narrowband modes, intercontinental QSOs only
1.840-1.850	CW, SSB, SSTV and other wideband modes, intercontinental QSOs only
1.850-2.000	CW, phone, SSTV and other wideband modes
80 Meters	
3.590	RTTY DX
3.610-3.630	RTTY
3.790-3800	DX window
3.845	SSTV
40 Meters	
7.040	RTTY DX
7.080-7.100	RTTY
7.171	SSTV
30 Meters	
10.140-10.150	RTTY
20 Meters	
14.070-14.099.5	RTTY
14.100	NCDXF beacons
14.230	SSTV
17 Meters	
18.100-18.110	RTTY
15 Meters	
21.070-21.100	RTTY
21.340	SSTV
12 Meters	
24.920-24.390	RTTY
10 Meters	
28.070-28.150	RTTY
28.190-28.225	Beacons
28.680	SSTV
29.300-29.510	Satellite downlinks
29.520-29.580	Repeater inputs
29.600	FM simplex
29.620-29.680	Repeater outputs

ARRL band plans for frequencies above 28.300 MHz are shown in the *ARRL Repeater Directory* and *FCC Rule Book.* For suggested packet frequencies, see *QST,* Sep 1987, p 54 and Mar 1988, p 51.

to slow down to 15 or 20 WPM for your benefit is asking quite a bit. Most amateurs enjoy working CW at or around a particular word-per-minute rate.

Keep your sending and receiving abilities on a par with one another. Being able to send at 30 WPM is of little value if you are only able to receive at 15. And although the ability to receive at a higher rate of speed is an admirable goal to pursue, you must be certain when sending at the same rate that you are able to do so with a skilled fist. To do otherwise will leave a poor impression on the recipient of your message.

Be courteous, to both the operator you are working and to others on the band. Avoid frequencies commonly used by nets or other operations such as W1AW code practice. Of course you must remain within the frequencies allowed for amateur CW use as specified by the FCC, but also try to stay within the ARRL's "Considerate Operator's Guide" of frequencies. This guide is a recommended spectrum usage designed to accommodate various amateur interests. For example, on 2 meters there are those who are interested in moonbounce work, and there is a portion of the band specifically set aside for that purpose. This is true for FM work, SSB and CW.

One example that stands out in the author's mind as a good illustration of courteous operation, or lack thereof, occurs when an amateur wishes to end a QSO for any one of a variety of reasons. It is quite common to hear "Well OM, must go now. XYL just rang the dinner bell," or "Must sign for now. Have to take daughter to school," or "Have to go. Getting up early tomorrow for work." Too often, in cases like these, the next thing that can be heard is a call or calls from other stations wishing to contact the station about to leave the air. Perhaps he is a rare DX find or new grid square for those trying to reach him. But regardless of why others may be calling, courtesy dictates that they should not do so. If someone must leave the air—let him go! Show him the consideration you would expect yourself.

Last, but by no means least, make sure your transmitted signal is clean — free of harmonics, key clicks, chirps or hum. An unclean signal always leaves a poor impression, no matter how good the operator is behind the key.

Break-in

Break-in is a type of CW operation in which the sender is able to listen between Morse characters or portions of characters while transmitting.

Semi break-in means that reception is possible between something other than character elements, such as between characters themselves. The actual delay between the time transmission stops and reception begins is usually variable. For example, for the same delay setting reception may be allowed between characters when sending at 20 WPM but only between words when transmitting at 40 WPM. Full break-in allows reception between character elements.

There are many benefits to break-in operation. For example, the length of time that a CQ must be sent can be reduced. As soon as a CQ is heard and a station is ready to respond, he can indicate this fact by "breaking in" on the CQ as it is being sent. If both stations in a QSO are using break-in they can respond to each other immediately if necessary. This may be done to indicate that a certain portion of the other's transmission was missed, or to insert appropriate comments during a QSO for any number of reasons. It is easy to see how the speed and ease of operation can be increased dramatically with break-in use.

The ability to use break-in can be indicated to other stations by use of the QSK Q-signal (I am equipped for break-in operation). This may be sent during a CQ or during the initial contact with another. Alternately, the BK symbol may be used while sending CQ, such as

CQ CQ CQ CQ DE W1AAA W1AAA BK CQ CQ...

CHAPTER SEVEN

Distress Calls

With 50,000 registered horsepower, 46,300 gross tons, and 92 feet wide, 175 feet high, almost three football fields in length and unsinkable—the *Titanic* cuts through the ice cold waters at latitude 41°46′ N, longitude 50°14′ W, on April 14, 1912. She is secure in her knowledge that her double bottom and 16 watertight compartments will keep her safe from harm. She was engineered to stay afloat with any two of the 16 compartments flooded, and it was impossible to imagine anything worse than that happening.

On deck, Lookout Reginald Lee had been warned to watch for icebergs. The *America*, *Baltic*, *Californian*, *Caronia* and *Mesaba* had each reported sightings of ice in the area.

At 11:40 PM, Lee sees the outline of an obscure figure in the darkness. At first it appears rather small, but then it grows larger and larger as the ship continues to steam westward. Suddenly, he realizes what he has seen and frantically calls the bridge. The bridge acknowledges, but for the next 37 seconds does little about the warning. Only during the last few seconds before impact does the *Titanic* veer to port, narrowly averting a head-on crash with an iceberg that looms 100 feet above the surface of the sea.

The resulting brush with the berg was hardly noticed. Some described the encounter with the huge chunk of ice as feeling like the ship had just nudged into dock. Another guessed the ship had dropped a propeller. Others assumed it was a rough sea.

After Captain Smith learned what happened, he ordered an inspection of the ship. Shortly thereafter, he ordered Radio Operator Jack Phillips to send out a call for help. It was 12:15 AM,

April 15, 1912. Phillips grasped his telegraph key and began. CQD MGY, CQD MGY, CQD MGY. CQD was the international distress call used at the time, and MGY were the call letters of the *Titanic*.

Again and again he called. Signals from the great ship began to be heard around the world. Charles B. Ellsworth, a 17-year-old wireless operator at Cape Race, Newfoundland, was one of the first to receive the unbelievable message. David Sarnoff, a young operator stationed on top of Wanamaker's Department Store in New York City, also intercepted the call. The *Carpathia*, 58 miles from the *Titanic*, heard the call and steamed full speed to her rescue.

Captain Smith had an idea. In an attempt to raise more help for his ship, he ordered Operator Phillips to send a different distress call. Although CQD was the traditional call, a recent international convention had agreed to use a new distress call—$\overline{\text{SOS}}$. And so Phillips tried.

The *Carpathia*, nearest liner to hear *Titanic*'s call, reached her first. Other ships also sped to her aid—the *Baltic*, *Olympic*, *Virginian*, *Parisian* and others. The rest of the story is well-documented history. Wireless saved the lives of 712 persons. The tragic part of the story—1517 lives were lost. And perhaps the greatest tragedy of all—the *Californian* was only 10 miles away when the first call for help was transmitted. She never heard *Titanic*'s plea because her radio operator was not on duty at the time.

The story of the *Titanic* is only one in a long list of sea disasters that have been reported by wireless. There were others before her and many following. There was the *Goodwin Sands* (April 28, 1899), the *Republic* (January 23, 1909), the *Delhi* (December 13, 1911), the *Veronese* (January 1, 1913), the *Templemore* (September 30, 1913) and many many more. And ships at sea are only one part in the telegraphy distress-call drama. Morse code has been used to summon help during air disasters, hurricanes, floods and wars.

But what does all this mean to you today? Where might you hear a distress call? How do you identify one? What do you do if you hear such a call? Where do you report the incident?

Handling Distress Calls

A distress call can be received on any frequency. There are certain, designated commercial emergency calling frequencies. These are listed here for those who may be interested in monitoring them. But bear in mind these are just *suggested* frequencies. A station in distress will use any means at its disposal to announce its plight. This can very well mean transmitting on a frequency other than those listed.

Designated Emergency Frequencies

500 kHz —International code distress and calling.
2182 kHz —International voice distress, safety and calling.
4125 kHz —Supplementary distress frequency.
6215.5 kHz —Supplementary distress frequency.
8364 kHz —International code lifeboat, life raft and survival craft.
40.5 MHz —US Army FM distress.
121.5 MHz —International voice aeronautical emergency.
156.8 MHz —FM international voice distress and international voice safety and calling.
243.0 MHz —Joint/Combined military voice aeronautical emergency and international survival craft.

$\overline{\text{SOS}}$ is used to indicate a distress situation by Morse code. MAYDAY is used to indicate the same by voice. If you hear either of these, begin by *remaining calm*. Assume the call is a legitimate one because there are severe penalties for perpetrating a hoax of this nature. Then, gather as much pertinent information as possible, such as:

1) Call letters of the transmitting station.
2) Identification of the ship or airplane if a sea or air emergency.
3) Nature of the distress situation.
4) Exact location of those in need of assistance.
5) Time you heard the call.
6) Frequency on which the call was received.
7) Any specific needs or requests that might have been indicated by the calling station.

Once you obtain all possible information, report the situation immediately to the appropriate authorities. Many agencies handle emergencies within the United States and its bordering waters, but the largest and best equipped organizations are the US Air Force and US Coast Guard. Each of these has facilities for coordinating search and rescue (SAR) operations with one another and with local authorities. The Air Force is responsible for Inland SAR and the Coast Guard for Maritime SAR. The Civil Air Patrol helps the Air Force.

Air Force SAR operations are coordinated through the Air Force Rescue Coordination Center (AFRCC) at Langley Air Force Base in Virginia. An emergency situation within the continental United States should be reported to your local Federal Aviation Administration (FAA) office, which will then notify the Air Force. If, for some reason, you are unable to reach your local FAA office, the Air Force SAR center can be reached directly by calling toll-free 1-800-851-3051. This number is to be used for emergency calls *only.*

Coast Guard efforts are coordinated through one of their Rescue Coordination Centers (RCC), which are the hub of operations for various areas along the Atlantic and Pacific Coasts. The Coast Guard has divided the Atlantic and Pacific regions into subregions and sectors as follows. Each of these subregions and sectors has its own RCC.

Atlantic Maritime Region

Boston subregion
Norfolk subregion
Miami subregion
New Orleans subregion (including Mississippi, Missouri, and Ohio Rivers)
Cleveland subregion (Great Lakes)
Western Atlantic Oceanic subregion (see map)

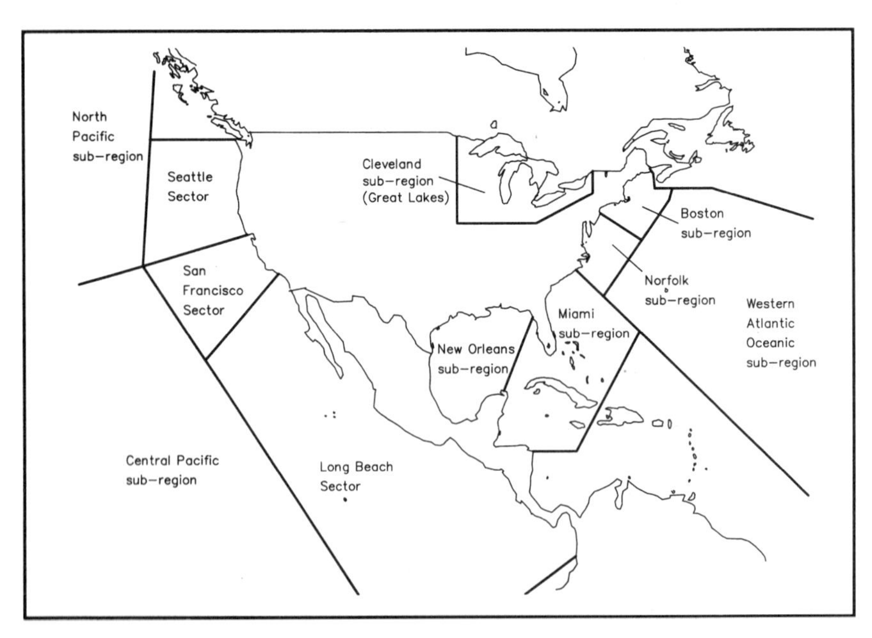

Pacific Maritime Region

Northern Pacific subregion
Central Pacific subregion
East Pacific subregion
Seattle sector

An emergency situation along the Atlantic or Pacific coast should be reported to the appropriate Coast Guard RCC. If you are not sure which subregion or sector has jurisdiction over the situation, call your nearest Coast Guard facility and they will contact the proper center.

Additional Information About Distress Signals and Search and Rescue

The foregoing is all you need know to successfully report any distress message. However, it is interesting to learn more about distress signaling and SAR operations.

For example, in the first six and one-half years of operation, the Air Force Rescue Coordination Center investigated 67,521 incidents and conducted 11,581 SAR missions that resulted in 5318 lives saved. This organization is only one part of a larger Air Force recovery team known as the Aerospace Rescue and Recovery Service (ARRS), which has saved the lives of more than 20,000 people in distress since its inception.

Certain frequencies have been specified for use during SAR missions. These are listed here for those interested in monitoring SAR operations.

3023 kHz —International voice SAR on scene.
5680 kHz —International voice SAR on scene.
123.1 MHz —International voice SAR on scene.
138.78 MHz —US military voice SAR on scene and direction finding.
155.16 MHz —FM frequency for on scene SAR operations coordination between states, local agencies and federal SAR services.
282.8 MHz —Joint/Combined on scene and direction finding.

Many calls for assistance originate from Emergency Locator Transmitters (ELTs) and Emergency Position-Indicating Radio

Beacons (EPIRBs). ELTs are carried by aircraft and EPIRBs are carried by ships, but each performs the same function. Both emit a signal on 121.5, 243 or 406.025 MHz when a distress situation is encountered.

For years, these units have been used to save the lives of thousands involved in airplane and ship disasters. Until recently, only 121.5 and 243 MHz signals were used. These transmissions were usually picked up by aircraft flying overhead. Since signals at these frequencies are primarily line-of-sight and low power is used, it was difficult for a ship to hear the call from such a device. At sea level, a broadcast from an ELT or EPIRB could only be heard for a distance of perhaps three or four miles. Detection by aircraft also had its limitations; it relied on the chance that an aircraft may be flying within range and might hear the signal. But then, in 1983, a great step forward was taken in the ability to detect these emergency beacons.

In January of that year, an international effort of the United States, Canada and France created a Search and Rescue Satellite-Aided Tracking (SARSAT) system. The Soviet Union also participates in that program with their own COSPAS spacecraft. Hence the name SARSAT-COSPAS has been given to this combined effort. The SARSAT-COSPAS concept involves the use of multiple satellites in low, near-polar orbits listening for distress signals.

The frequencies 121.5, 243 and 406.025 MHz are monitored. The first two frequencies, as mentioned earlier, had been in use for years. The latter was put into use just for satellite systems. Earlier experiments performed by the National Oceanic and Atmospheric Administration (NOAA) showed that higher frequencies provided superior Doppler shift tracking results. NOAA had used 401-MHz transmitters attached to icebergs, polar bears, sea turtles and balloons in their satellite-tracking experiments. Use of the higher frequency allowed greater precision in the location of tracked objects.

When a distress signal is received by one of the new SARSAT-COSPAS satellites, it is rebroadcast to a network of ground-receiving stations. The location of the emergency is determined by measuring signal frequency shifts, or Doppler shifts, as the satellite passes overhead. The information so received is then relayed to the Air Force Mission Control Center at Scott Air Force Base. Determination of the location of the ship or plane in distress is accurate to within 5 to 10 miles when a frequency of 121.5 MHz is used, or 2 to 5 miles when

406.025 MHz is used. Other benefits are derived from the use of the new 406.025 MHz signal. Coded into the beacon at this new frequency is the identification of the craft, nationality, classification (aircraft or ship), nature of the distress situation and elapsed time since the incident occurred.

Where to Report a Distress Call

For an Inland Emergency:

1) Call your local FAA office.

2) If unable to reach an FAA office, call the Air Force Rescue Coordination Center directly at 1-800-851-3051. This number is to be used for emergency calls *only*.

For an Atlantic or Pacific Emergency:
Notify the appropriate (or any) Coast Guard center below.

Atlantic Maritime Region

Boston subregion	617-223-8555
Norfolk subregion	804-398-6231
Miami subregion	305-536-5611
New Orleans subregion (including Mississippi, Missouri and Ohio Rivers)	504-589-6225
Cleveland subregion (Great Lakes)	216-522-3984
Western Atlantic Oceanic subregion (see map)	804-398-6231

Pacific Maritime Region	510-437-3709
Northern Pacific subregion	907-463-2000
Central Pacific subregion	808-541-2500
Eastern Pacific subregion	415-437-3700
Seattle sector	206-220-7001

CHAPTER EIGHT

Advances in Morse Technology

As with other communications modes, the means of sending and receiving Morse code are constantly under experimentation. With computer technology appearing in more and more ham radio equipment, the sophistication of amateur gear has increased dramatically. Let's look at some of the areas of Morse technology where advances are being made.

Coherent CW

Anyone who listens to today's ham bands knows that at peak conditions overcrowding often is a problem. As a result, anything that reduces the bandwidth of a transmitted signal, without damaging the information contained therein, is highly desirable. The inception of single sideband did this years ago for our phone bands.

Modern receivers, even those of high quality, typically use a 500- to 2000-Hz bandwidth for CW reception, although only 100 Hz is needed. This is the case for a variety of reasons. First, it is difficult mechanically and electrically to hold receiver and transmitter stability down to less than 100 Hz. Also, when bandwidths under several hundred hertz are used for reception, signals tend to "ring," thereby distorting the code. Both problems have been addressed as a result of advances in electronics.

Coherent CW (CCW) makes use of a single, highly accurate frequency source, such as WWV, as the basis for synthesizing all local oscillator frequencies needed in both receiver and transmitter. Then, by taking advantage of the precise time signals on this same frequency source, the code received by one station is matched, or synchronized, to that transmitted by another station (as long as the exact code speed is known by both stations). By so doing, a "matched" filter of very narrow bandwidth can be employed at the receiving end. Or, to put it more simply, the receiving end uses the same frequency and time standard as the transmitting end and, as a result, it "knows" a great deal about the transmitted signal before it even arrives. Therefore, receiving functions can be optimized accordingly. More than 20 dB improvement in signal strength can be realized by the use of this technique. The technicalities behind this process are beyond the scope of this book, but the reader is referred to a series of articles in May and June 1981 *QST*, which detail the theory and practice of CCW equipment.

Super CW

Anything that improves the reliability of communications is welcome and Super CW shows promise of doing just that. This communications concept, proposed by WØXI, makes use of computers on the receiving and transmitting ends of a two-way Morse code contact. One of three different modes called SS CW, FEC CW, and ARQ CW are used. All employ a fixed, high-speed rate of transmission.

In Standard Speed CW (SS CW), the transmitting station simply sends at a rate of speed that is known to the receiving station. Because the receiving computer is aware of the speed used, it can detect the code with greater accuracy. This is a somewhat similar concept to CCW discussed previously except that a common signal source is not used to synchronize transmission and reception. (CCW is synchronous; SS CW is asynchronous.)

FEC CW stands for Forward Error Correction CW. In this mode, each word is transmitted twice. If the computer at the receiving end detects both words without difficulty, it displays only one. If it receives one word but the other is damaged because of noise or interference, it still displays only one — the correct word. Through use of this method, communications reliability is greatly improved, but if difficulty is still encountered, ARQ CW can be used.

ARQ CW is an acronym for Automatic Repetition reQuest CW. Using this concept, the transmitting station sends each word twice, as with FEC CW. The receiving station again displays a single word if either one, or both, are received correctly. But then, it acknowledges correct reception by sending an "R" back to the transmitting station.

If no R is sent, the originating station retransmits the same word twice. This sequence is repeated up to four times if necessary.

Narrowband Filters

Narrow filter techniques have been around for years. They all have the same purpose, to reduce the width of the signal being received in order to eliminate adjacent channel interference and noise. Old-timers will remember the "Select-O-Ject" and "Q-Multiplier." And, of course, crystal filters have been a mainstay of CW operators for a long time.

Today, "Passband Tuning" is popular on many receivers. Passband Tuning allows the operator to electronically vary the bandwidth of the received signal in the intermediate frequency (IF) stage of a receiver, thereby accomplishing the same feat that required a separate external unit years ago. The frequency-width variation achieved is commonly between 2.2 kHz and 500 Hz.

Just as the passband may be narrowed in the IF section of a receiver, so may it be in the audio section. Audio filters that accomplish this feat were discussed previously. But there are different types (and prices) of audio filters. A state-of-the-art unit that deserves particular mention is the switched capacitor audio filter, sometimes called a SCAF.

Modular Systems (MSC) produces a SCAF that they call a "SMART" filter. A microcomputer within the unit is used to control complex operations of the filter so that the operator need only be concerned with adjusting the center frequency and bandwidth. Bandwidth is adjustable down to a surprisingly narrow 25 Hz. Extremely

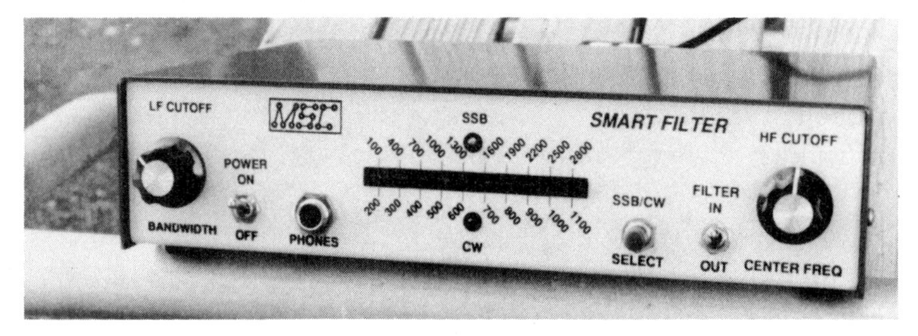

MSC SMART Filter

steep sides, or "skirts" on the cutoff frequency are another advantage of this type of filter.

Code Generators/Keyers

Random code generators were discussed in chapter 4. These devices have come a long way in recent years, also. Today, some not only generate code for training purposes, but have the ability to do a lot more. An example of one modern day unit is the AEA "Morse Machine."

This device can generate random code for training purposes, produce random 4-letter words, and has a QSO simulator called Dr. QSO. With it, you can call other stations, answer a CQ, or simply listen to realistic, on-the-air type QSO's.

The unit can also be used as an automatic keyer for calling CQ, contesting, beacon or other use. Morse speed is adjustable from 2 to 99 WPM. Twenty memories are provided that are capable of holding up to 8,000 characters of information (expandable to 36,000 characters). A lithium battery backup is used to maintain memory status when the unit is off.

A computer can be interfaced with The Morse Machine and any of its front panel functions may be programmed by the computer.

AEA Morse Machine

Code generated by the machine may also be displayed on the computer's CRT.

Quite a machine!

Computer Software

Computer software is being used by the amateur community in a variety of ways. In addition to its use for teaching purposes, software is now available to aid the amateur with almost every facet of his station's operation.

Examples of capabilities of some of the packages on the market include the ability to:

- Automatically log call sign, band, frequency, time and date of QSO.
- Search logged data by any field.
- Store and access notes on QSO's.
- Store all or part of any QSO on disk.
- Allow keyboard entry of frequency and mode.
- Provide assistance with awards tracking and contesting.
- Automatically determine beam headings.
- Print QSL cards, labels and reports.

Microsystems Software, Inc. provides an unusual software package for the CW enthusiast. Called HandyCODE AR, this package uses Morse code output to control a computer. It does this by translating Morse signals into the equivalent of keyboard keystrokes. This is useful for CW bulletin boards and other applications. The program also uses non-standard Morse characters to provide user-definable functions and macros.

Included is the capability for straight key, bug or keyer operation, a user-configurable iambic dot memory, and selectable code speeds from 1 to 99 WPM. In addition, HandyCODE AR is designed to permit operation by handicapped individuals.

CHAPTER NINE

Other Versions of the Code

Codes of Other Countries

Occasionally you may hear transmissions being sent in an unrecognizable code. These are often contacts between individuals of other nations conversing in their native language. The Morse code equivalents for some major foreign countries are presented here. It is interesting to note that the Soviet military, and the Navy in particular, make extensive use of CW. This may be due in part to the realization that code may prove to be the only successful means of communication when vulnerable satellites are destroyed or the effects of nuclear war render other methods unusable.

Dashless Version of the Code

There are times when it is difficult or impossible to make a *dah* when sending the code due to the fact that the means of transmission are limited. Examples would include attempting to transmit by tapping one object with another, such as striking two rocks together, tapping one's knuckles on a table and so forth. Perhaps no situation portrays restricted means of transmission better than a prisoner-of-war scenario in which men have need for secret, non-verbal communication, but do not have at their disposal the means to create a *dah* in the Morse code.

As a result of the above a dashless version of the code was put into use by prisoners-of-war in Vietnam. Known as the "Smitty Harris Tap Code," named after a prisoner-of-war who came across it by accident during survival school training, it became a standard for use by American POW's in need of a code of this type. The story behind its development and use can be found in a book entitled *In Love and War* by Jim and Sybil Stockdale, the story of a family's ordeal and sacrifice during the Vietnam War, alternately published by Harper and Row (1984), Bantam Books (1985) and Naval Institute Press (1990). Since the code is rather laborious to send, abbreviations were frequently used. Note that the C and K are identical.

Morse Codes for Other Languages

Code	*Japanese*	*Thai*	*Korean*	*Arabic*	*Hebrew*	*Russian*	*Greek*
•	ヘ he	เ sara-a	ㅏ a		ו vav	Е,э E	Ε epsilon
–	ム mu	ต tor-tow	ㅓ ŏ	ت ta	תת tav	Т T	Τ tau
••	゛ nigori	ิ sara-e	ㅑ ya	ي ya	י yod	И I	Ι iota
•–	イ i	า sara-r	ㅗ o	ا alif	א aleph	А A	Α alpha
–•	タ ta	น nor-nu	ㅛ yo	ن noon	נן nun	Н N	Ν nu
––	ヨ yo	ม mor-ma	ㅁ m	م meem	מם mem	М M	Μ mu
•••	ラ ra	ส sor-sue	ㅕ yŏ	س seen	שש shin	С S	Σ sigma
••–	ウ u	่ mai-ek		ط ta	ט tet	У U	ΟΥ omicron ypsilon
•–•	ナ na	ร roe-rue	ㅠ yu	ر ra	ר reish	Р R	Ρ rho
•––	ヤ ya	ว vor-van	ㅂ p(b)	و waw	צץ tzadi	В V	Ω omega
–••	ホ ho	ด door-dek		د dal	ד dalet	Д D	Δ delta
–•–	ワ wa	ค kor-kwai	ㅇ -ng	ك kaf	ככך chaf	К K	Κ kappa
––•	リ ri	ก go-kai	ㅅ s	غ ghain	ג gimmel	Г G	Γ gamma
–––	レ re	โ sara-o	ㅍ p′	خ kha	ה heh	О O	Ο omicron
••••	ヌ nu	ห hor-heep	ㅜ u	ح ha	ח chet	Х H	Η eta
•••–	ク ku	้ mai-tho	ㄹ r-(-l)	ض dad		Ж J	ΗΥ eta ypsilon
••–•	チ ti	ฟ for-fun	ㄴ n	ف fa	פף feh	Ф F	Φ phi
••––	ノ no	ึ sara-eu				Ю yu	ΑΥ alpha ypsilon
•–••	カ ka	ล law-ling	ㄱ k(g)	ل lam	ל lamed	Л L	Λ lambda
•–•–	ロ ro	แ sara-air		ع ain		Я ya	ΑΙ alpha iota
•––•	ツ tu	ป por-pla	ㅈ ch(j)		פ peh	П P	Π pi
•–––	ヲ wo	ญ yor-ying	ㅎ h	ج jeem	ע ayen	Й Y	ΥΙ ypsilon iota
–•••	ハ ha	บ bor-baimai	ㄷ t(d)	ب ba	בב bet	Б B	Β beta
–••–	マ ma	ช chor-chang	ㅋ k′	ص sad		Ъ,ь mute	Ξ xi
–•–•	ニ ni	ข kho-khai	ㅊ ch′	ث tha	ס samech	Ц TS	Θ theta
–•––	ケ ke	ย yor-yak	ㅔ e	ظ za		Ы I	Υ ypsilon
––••	フ hu	ซ zor-zo	ㅌ t′	ذ dhal	ז zain	З Z	Ζ zeta
––•–	ネ ne	ผ pho-phueng	ㅐ ae	ق qaf	ק kof	Щ SHCH	Ψ psi
–––•	ソ so	ุ sara-u		ز zay		Ч CH	ΕΥ epsilon ypsilon
––––	コ ko	ฉ choe-ching		ش sheen		Ш SH	Χ khi
••–••	ト to	ี sara-ie		ه he			
••–•–	ミ mi	ู sara-au					
••––•	゜ han-nigori	ื sara-aue					
•–•••	オ o	ะ sara-ar					
•–••–	ヰ (w)i	ไ sara-i					
•–•–•	ン n	๋ mai-jatawa					
•–•––	テ te	ฤ row-rue					
•––••	ヱ (w)e	พ por-pan					
•––•–	- hyphen	ั mai-han agas					
•–––•	セ se	็ maitho han agas					
–•••–	メ me	อ or-ang					
–••–•	モ mo	จ jor-jan					
–••––	ユ yu	ท tor-tahan					
–•–••	キ ki	ถ tor-tung					
–•–•–	サ sa						
–•––•	ル ru	ง ngor-ngoo					
–•–––	エ e						
––••–	ヒ hi						
––•–•	シ si						
––•––	ア a	ฮ hor-nok hook					
–––•–	ス su						
•–•••–				لا lam-alif			

The Smitty Harris Tap Code

A . .	B . ..	C/K	D	E . .....
F .. .	G	H	I	J .. ..
L	M	N	O	P ... .
Q	R	S	T	U ....
V	W	X	Y	Z ....

CHAPTER TEN

Compendium of CW-Related Articles

THE CODE, GENERAL

"**Some Thoughts on the Morse Code**," Victor C. Clark, W4KFC
QST, December 1982, page 55
A well-thought-out, objective look at the code.

"**Novice Notes: The Case for Code**," Rick Booth, KM1G
QST, July 1989, pages 47-48
Reasons why Novices may very well prefer CW over phone operation.

"**Why CW?**" Doug Stivison, WA1KWJ
Ham Radio Horizons, February 1978, pages 42-46
More reasons why many amateurs prefer CW over all other modes of communication.

"**The K7DBV Guide to Easy CW QSO's**," Gene A. Williamson, K7DBV
QST, August 1986, pages 48-49
Proper CW operating practices.

"**How to Increase Your QSO's**," Leon Fletcher, N6HYK
73 Amateur Radio, September 1983, pages 34-38
Advice on handling CW contacts.

"**Radiotelegraph Codes: There's not Just One**," W. Clem Small, KR6A
Ham Radio, September 1988, pages 82-83
Dissertation on the different types of radiotelegraph codes used in the early 1900's.

"**CW Anyone?**" Harry W. Lewis, W7JWJ
Ham Radio, March 1981, pages 44-45
Information on CW world's speed records.

"**Worldwide CW Codes**," Bill Welsh, W6DDB
CQ, December 1980, pages 32-35
Codes of other nations.

"**Plenty of Morse Here!**"
Morsum Magnificat, Spring 1988 (no. 7), page 31
Comments on the extensive use of code by the Soviets.

"**It's All Morse!**" Lewis Coe, W9CNY
CQ, January 1989, pages 52-54
Discussion of the code from its invention up to modern day use.

"**The Morse Code: Alive and Kicking,**" Karl Thurber
Popular Electronics, April 1981, pages 89-90
General discussion on CW and its usefulness.

"**Morse Code — The International Language,**" Mike Mitchell, Jr., W7WHT
Monitoring Times, October 1987, pages 30-31
About the code, its history and techniques of learning.

"**Did Morse Get It Right?**" A. S. Chester
Morsum Magnificat, Spring 1987 (no. 3), pages 18-26
An excellent study of the code, analyzing the definition of a "mean word" in Morse telegraphy as it is used to arrive at a words-per-minute speed figure.

"**Q and Z Codes,**" Gerald Stancey, G3MCK
Morsum Magnificat, Autumn 1990 (no. 17), page 13
Comments on the amateur's use of Q and Z codes.

"**The Hebrew Telegraph Code,**" Donald K. de Neuf, WA1SPM
Morsum Magnificat, Autumn 1990 (no. 17), page 39
History of the Hebrew telegraph code.

LEARNING THE CODE

"**Learning Morse,**" Vincent O'Keeffe, WA1FKF
QST, August 1972, pages 58-62, 70
An excellent study on learning the Morse code. Contains many references from various psychological studies.

"**Learning Code by Osmosis,**" George Hart, W1NJM
QST, August 1979, pages 58-59
Some common sense advice on learning the code.

"**Morse Decoded,**" Charles Harris, WB2CHO
QST, September 1976, pages 53-54
Discussion on the basics of learning the code and what's available to aid you in the process.

"**A Fresh Look at CW,**" Don Daso, WA8MAZ
QST, August 1987, pages 44-47
Advice for the newcomer on gaining CW operating proficiency.

"**High-Speed CW, Anyone?**" George Hart, W1NJM
QST, June 1979, pages 51-52
Story of the on-the-air, high-speed code program of W1NJM.

“**Sending!**” William G. Gerlach, W6BG
QST, November 1974, page 75, 80
Discussion of proper sending habits using a bug.

“**Throw Away Your Pencil!**” Melvin Broaddus, K6LJE
QST, April 1986, pages 43-44
Copying code in your head.

“**How to Improve Your Code Speed**,” Edward F. Rice, W9NGP
73 Amateur Radio, July 1988, page 21

“**Morse Code Motivational Techniques**,” Peter Kemp, KZ1Z
73 Amateur Radio, July 1988, page 24

“**Problems in Learning the Code**,” Larry Lisle, K9KZT
73 Amateur Radio, July 1988, pages 54-55
Tips for code class teachers.

“**Increase CW Speed**,” Mayer D. Zimmerman, W3GXK
CQ, October 1987, page 106

“**So You Want to Increase Your Code Speed**,” Arthur R. Lee, WF6P
CQ, December 1987, pages 56-57

“**Bill’s Basics: Code and Code Receiving Practice**,” Bill Welsh, W6DDB
CQ, November 1989, pages 86-88

“**Learning the Morse Code — Part 1**,” Bill Welsh, W6DDB
CQ, June 1979, pages 58-59

“**Learning the Morse Code — Part 2**,” Bill Welsh, W6DDB
CQ, July 1979, pages 48-49

“**Learning the Morse Code — Part 3**,” Bill Welsh, W6DDB
CQ, August 1979, pages 51-54

“**One Last Crack at the Code**,” David Kaufman, WA3WBI
CQ, August 1978, pages 26-28
A learning method.

“**Some Ideas on Code Practice**,” Solomon Kupferman, W2GVT
CQ, November 1975, pages 37-38, 79-80

“**Novice Code Problems**,” Herbert S. Brier, W9EGQ
CQ, January 1974, pages 51-53

“**Phantom Copying — A Transitional Step to QRQ**,” Al D’onofrio, W2PRO
CQ, June 1974, pages 21-24, 69-72
Mastering the technique of copying behind and copying in your head.

“**Tips for Copying CW on Paper**,” Al D’onofrio, W2PRO
CQ, November 1973, pages 38-39

“**The Keyer is the Key**,” David Kaufman, WA3WBI
CQ, July 1979, pages 58-59
Developing good sending habits.

"**Drills to Build Skill in Copying Behind**," William G. Pierpont, NØHFF
Morsum Magnificat, Summer 1989 (no. 12), pages 18-20
Exercises to help one learn the difficult skill of recording Morse behind the text currently being sent.

"**Morse Testing and Training in the UK**," Ron Wilson, G4NZU
Morsum Magnificat, Summer 1988 (no. 8), pages 22-27
Morse learning and licensing practices in the United Kingdom.

"**Walter Candler and the Candler System**," William G. Pierpont, NØHFF
Morsum Magnificat, Winter 1988 (no. 10), pages 1-6
Story of the once popular Candler System of learning the code.

"**How to Break the 10-Words-Per-Minute Code Barrier**," Richard Humphrey
Popular Electronics, July 1974, pages 67-68

"**Some Notes on Acquiring the Code**," John B. Johnston, K3BNS
QST, November 1966, pages 62-63
Notes, tips and techniques on improving code proficiency.

"**Butchering the Code**," William G. Pierpont, NØHFF
Morsum Magnificat, Spring 1987 (no. 3), pages 9-11
Discussion of proper sending habits and drills to improve same.

"**You Will Upgrade!**", David Schoenthaler, KØHBQ
73 Amateur Radio, July 1988, pages 34-36
Hints on passing the code test.

TELEGRAPH KEYS, GENERAL

"**Telegraph Keys: As American as Pumpkin Pie**," Joel P. Kleinman, WA1ZUY
QST, December 1979, pages 18-19

"**Return of the Classic Keys**," Dave Ingram, K4TWJ
CQ, June 1987, pages 62-68

"**More Classic Keys and Telegraph Tales**," Dave Ingram, K4TWJ
CQ, February 1988, pages 72-75

"**The Story of the Key — Part 1**," Louise Ramsey Moreau, W3WRE
Morsum Magnificat, Winter 1987 (no. 6), pages 1-13
This is the first of a 6-part series of excellent articles on the telegraph key by an expert on the subject.

"**The Story of the Key — Part 2**," Louise Ramsey Moreau, W3WRE
Morsum Magnificat, Spring 1988 (no. 7), pages 13-20

"**The Story of the Key — Part 3**," Louise Ramsey Moreau, W3WRE
Morsum Magnificat, Summer 1988 (no. 8), pages 1-9

"**The Story of the Key — Part 4**," Louise Ramsey Moreau, W3WRE
Morsum Magnificat, Autumn 1988 (no. 9), pages 35-41

"**The Story of the Key — Part 5**," Louise Ramsey Moreau, W3WRE
Morsum Magnificat, Winter 1988 (no. 10), pages 31-39

"**The Story of the Key — Part 6**," Louise Ramsey Moreau, W3WRE
Morsum Magnificat, Spring 1989 (no. 11), pages 26-34

"**Key Collecting**," Colin Waters, G3TSS
Morsum Magnificat, Winter 1989 (no. 14), pages 12-14

"**The W7GAO Key Collection**," Martin W. Krey, K7NZA
73 Amateur Radio, May 1979, pages 38-42

"**A Tour of the K5RW Key Collection/Museum — Part 1**," Dave Ingram, K4TWJ
CQ, May 1985, pages 82-84

"**A Tour of the K5RW Key Collection/Museum — Part 2**," Dave Ingram, K4TWJ
CQ, June 1985, pages 80-84

"**Historic Key of 1AW**," Tony Smith, G4FAI
Morsum Magnificat, Autumn 1989 (no. 13), pages 40-41
The story of the telegraph key of Hiram Percy Maxim, co-founder of the American Radio Relay League.

"**Eddystone Bug**," Colin Waters, G3TSS
Morsum Magnificat, Autumn, 1989 (no. 13), pages 30-33

"**The Double Speed Key**,"
Morsum Magnificat, Summer 1989 (no. 12), pages 27-30
An excellent article on the "sideswipe" or "Cootie key."

"**Australian Jiggers**," John Houlder
Morsum Magnificat, Spring 1989 (no. 11), pages 21-23
A discussion of Australian landline bugs, known as "jiggers."

"**World of Ideas: More Keys Revisited**," Dave Ingram, K4TWJ
CQ, February 1989, pages 64-70
Photographs and stories of unusual telegraph keys.

"**The Iambic Gambit**," Lew Fay, AA5Q
QST, July 1981, page 52
A discussion of iambic keying techniques.

"**Keys, Keyers and Keyboards**," Bruce Hale, KB1MW
QST, December 1989, pages 47-49
A general discussion of straight keys, bugs, keyers, iambic keyers, memory keyers and keyboards.

"**Showcase**"
Morsum Magnificat, Autumn 1986 (no. 1), pages 35-38
also: Winter 1986 (no. 2), pages 18-21
Spring 1987 (no. 3), pages 2-23
Summer 1987 (no. 4), pages 17-20
Autumn 1987 (no. 5), pages 42-44
Winter 1989 (no. 14), pages 36-39
Spring 1990 (no. 15), pages 26-28
Pictures and descriptions of unusual telegraph keys.

"**World of Ideas: Classically Classic Keys**," Dave Ingram, K4TWJ
CQ, May 1990, pages 108-113
Pictures and descriptions of unusual telegraph keys.

"**World of Ideas: A Keys Bonanza!**", Dave Ingram, K4TWJ
CQ, November 1990, pages 79-86
Photographs of and stories behind a variety of telegraph keys.

"**G4ZPY Keys**"
Morsum Magnificat, Autumn 1990 (no. 17), pages 36-38
Story behind one of Britain's newest telegraph key manufacturers.

"**Neoprene Foam Aids Keyer-Paddle Stability**" (A *Hints and Kinks* item), David Frost, VE7FJE
QST, August 1988, page 50

"**Right- or Left-Hand Paddle Operation**" (A *Hints and Kinks* item), Robert L. Vandevender II, KR2K
QST, June 1987, page 40
Simple technique for right- and left-handed keyer-paddle operation without making changes to the paddle itself.

"**On the Ham Bands: Go Pound Brass**," Ike Kerschner, N3IK
Monitoring Times, July 1989, pages 44-45
Discussion of telegraph keys and keyers from early straight keys to modern day keyboards.

TELEGRAPH KEYS, CARE AND ADJUSTMENT

"**How to Adjust a Straight Key and Send Good Code**," Jim Bartlett, WB9VAV
QST, December 1977, pages 15-17

"**Semi-Automatic Key Adjustment**," Brian Murphy, VE2AGQ
QST, February 1968, pages 60-61

"**Adjusting and Cleaning of Speed Keys (Bugs)**," Al D'onofrio, W2PRO
CQ, August 1972, pages 36-37, 92

"**Tuning Your Vibroplex**," Jens H. Nohns, OZ1CAR
Morsum Magnificat, Winter 1989 (no. 14), page 46

"**How to Improve the Vibroplex EK-1 'Brass Racer' Keyer,**"
Albert H. Jackson, VE3QQ
CQ, October 1984, pages 13-19

"**Bug Dot Damper,**" Noel T.J. Brown, GW8IH
Morsum Magnificat, Summer 1990 (no. 16), page 7
A dot contact spring damper to prevent scratchiness often encountered with bugs.

"**Cleaning Key Contacts**" (A *Hints and Kinks* item), Joe Rice, W4RHZ
QST, March 1988, page 40
How to clean key contacts without a special burnishing tool.

"**Slower Keying Speeds With A Bug**" (A *Hints and Kinks* item),
Edward Peter Swynar, VE3CUI
QST, June 1988, page 49
How to reduce bug keying speeds below normally adjustable levels.

TELEGRAPH KEYS, HOME-MADE

"**A Key to Success,**" Daniel Wissell, WB2MFH
QST, December 1977, pages 19-20
How to build your own straight key.

"**Zero-Cost Key,**" Antonio G. O. Gelineau, W1HHF
QST, November 1980, pages 32-33
A homemade straight key/paddle.

"**Daniel's Key,**" Daniel L. Nevels, WD5ETR
QST, December 1977, pages 21-22
A key that works on infra-red light.

"**Home-Made Key,**" Barrie E. Brokensha, ZS6AJY
Morsum Magnificat, Autumn 1989 (no. 13), pages 23-26

"**Brass Hand Key,**" Don Harris
Morsum Magnificat, Spring 1989 (no. 11), pages 10-11
A homemade, solid brass hand key.

"**Construct the Lead-Foot Key,**" J. H. Owens, W5JQE
73 Amateur Radio, July 1983, pages 48-49
A home-built keying paddle.

"**Home-made Handkey,**" Maurice Sandys, G3BGJ
Morsum Magnificat, Spring 1987 (no. 3), pages 12-17
A home-made straight key patterned after the Royal Air Force Type D, the standard ground station key used at the outbreak of World War II.

"**The Mini-Mouse Key,**" Erich A. Pfeiffer, WA6EGY
73 Amateur Radio, January 1979, pages 120-122
A homemade paddle for electronic keyers.

''**External Paddles for the Heath HD-1410 Keyer**,'' David L. Ashenfelter, KA8DDT
QST, October 1980, page 25

''**A Poor Boy's Paddle**,'' Howard Goldstein, WB2IWX
QST, December 1977, pages 22-23
A homemade keyer paddle.

''**The Stark Key**,'' Howard J. Stark, W4OHT
QST, May 1977, pages 33-34
A homemade keyer paddle.

''**How to Make a Low-Cost Keying Mechanism**,'' A. K. Weis, WA5VQC
QST, August 1971, pages 22-23
Converting a bug to a keyer paddle.

''**How to Build a Junkbox Paddle Key**,'' John R. Somers, KC3YB
CQ, February 1990, pages 34-36
A homemade iambic key.

''**An Iambic Key for the Heath HW-8**,'' F. K. Feeney, WB2EMS
73 Amateur Radio, November 1987, pages 32-33

''**Kaboom Micro Keyer**,'' Michael Jay Geier, KB1UM
73 Amateur Radio, September 1989, pages 28-29
A miniature iambic keyer.

''**How to Build Your Own Mini Hand Key**,'' George A. Wilson, Jr., W1OLP
CQ, June 1990, pages 34-35
A miniature key, only 3½ inches in length.

''**0073 Spy Key**'' (QRP Department), Skip Westrich, WB8OWM
73 Amateur Radio, October 1988, page 83
Very small, unusual straight key made from two dominoes and a Radio Shack micro-switch.

''**A Breath-Actuated Key**'' (A *Hints and Kinks* item), Julian S. Lorenz, MD, KE6VL
QST, March 1989, page 39
Very unique, homemade key that is actuated by the breath. This unit can be used as a novelty item or it can be a tool for the handicapped amateur.

''**A Homemade Keyer Paddle**'' (A *Hints and Kinks* item), Arnold Harvey, WB8OJN
QST, May 1984, pages 43-44

''**Keyer-Paddle Construction Ideas**'' (A *Hints and Kinks* item), Antonio G.O. Gelineau, W1HHF
QST, May 1984, page 44
Variety of ideas for making homemade keyer-paddles with simple hand tools.

CW KEYERS

"**The AD7Iambic Cheap Keyer**," Paul Newland, AD7I
QST, June 1988, pages 38-41
A four-IC CMOS keyer.

"**The UNKEMO (UNiversal KEyer MOdule)**," George Murphy, VE3ERP
QST, February 1986, pages 27-28, 50
An electronic keyer that works with any rig, regardless of whether negative, positive or grid-block keying is used.

"**CW on a Chip**," Bob Shriner, WAØUZO and Paul K. Pagel, N1FB
QST, December 1983, pages 16-19
A solid state iambic keyer.

"**The $2 Infinite-Memory Keyer**," Phil Anderson, WØXI
73 Amateur Radio, July 1983, pages 50-51

"**QRP Keyer for Misers**," T. K. Davies, VE7DHD
73 Amateur Radio, July 1983, pages 58-63

"**The Soft Touch Keyer**," Nelson Sollenberger, WA3PKU
73 Amateur Radio, January 1979, pages 128-130
An electronic keyer with no moving parts that operates from just the touch of the user's hand.

"**Sound-Sensitive CW Sender**," Len Furman, W2GZ
73 Amateur Radio, April 1980, page 106
This is a unique CW keyer for the manually handicapped amateur that keys a rig by voice operation.

"**Simple, Compact QRP Keyer**," Jack Najork, W5FG
Ham Radio, October 1984, pages 82-83

"**Cubic Inch Keyer**"
73 Amateur Radio, November 1987, pages 32-33
An iambic keyer for the Heath HW-8.

"**Digi-Keyer**," Ronald D. King, AB4DP
Ham Radio, October 1989, pages 68-71
An electronic keyer using TTL 7400 series digital IC's.

"**Keys, Keyers and Keyboards**," Bruce Hale, KB1MW
QST, December 1989, pages 47-49
A general discussion of straight keys, bugs, keyers, iambic keyers, memory keyers and keyboards.

"**Key-Tronics — Part 1**," Roy Walmsley, G3IBB
Morsum Magnificat, Spring 1987 (no. 3), pages 35-38
Discussion on electronic keyers, their use and shortcomings.

"**Key-Tronics — Part 2**," Roy Walmsley, G3IBB
Morsum Magnificat, Summer 1987 (no. 4), pages 42-43
How to set the weight control on your electronic keyer for best results.

"**Keyers With Automatic Spacing,**" Dr. Gary Bold, ZL1AN
Morsum Magnificat, Spring 1990 (no. 15), pages 12-14
About the auto-space feature on certain electronic keyers that enables the operator to send perfect code by automatically generating correct character spacing.

"**Paddle and Keyer Notes on Current Amateur CW Technique,**" Bob Locher, W9KNI
CQ, December 1986, pages 20-21
How to use iambic and non-iambic keyers.

"**All This and PCII,**" Bob Lockwood, W4FXI
73 Amateur Radio, March 1985, pages 26-27
Programmable Morse keyboard and memory keyer built from a Radio Shack/Sharp pocket computer.

"**DIY CW Keyer**" (QRP Department), Mike Bryce, WB8VGE
73 Amateur Radio, July 1988, pages 80-81
A CW keyer, not much larger than a 9 volt transistor radio battery, built around two IC's.

"**The CMOS Super Keyer II,**" Jeff Russell, KCØQ and Bud Southhard, NØII
QST, November 1990, pages 18-21
Improved version of CMOS Super Keyer that appeared in October, 1981 *QST.*

"**Avoiding Static Damage to the Heath μMatic Memory Keyer**"
(A *Hints and Kinks* item), John DeCicco, KB2ARU
QST, September 1989, page 37
Useful tip for preventing damage to this unit.

"**Fine Tune the Speed of your Yaesu FT-757X Transceiver's Internal Keyer**"
(A *Hints and Kinks* item), Roger Burch, WF4N
QST, June 1989, page 42
Modifying the 757X's keyer speed control for smoother operation.

"**Making the MFJ-484 Grand-Master Keyer a Bit Grander**" (A *Hints and Kinks* item), Jon Zaimes, AA1K
QST, April 1989, pages 39-40.
A series of worthwhile improvements for this keyer.

"**Keys and Keyers**" (A *Hints and Kinks* item), Robert G. Wheaton, W5XW
QST, April 1986, page 39
A variety of tips for keyer users.

"**Magnetic Switch for CW Tune Up**" (A *Hints and Kinks* item), Rick Lucas, WBØNQM
QST, August 1983, page 41
Way to provide key-down condition for keyers without a TUNE position.

CW KEYBOARDS

"**A Morse Keyer Using the Coleco Adam Keyboard**," Ed Oscarson, WA1TWX
QST, June 1989, pages 27-34

"**The CW6805 — An Inexpensive Morse Keyboard**," Ed Oscarson, WA1TWX
QST, December, 1988, pages 24-30
A keyboard built around the Motorola MC68705P3 single-chip microprocessor.

"**Send Error-Free Code With One Hand**," W. E. Quay, W4MKC and R. H. Turrin, W2IMU
QST, January 1986, pages 25-28
A one-hand keyboard keyer.

"**A Keyboard Keyer and Code-Practice System**," Dan Whipkey, N3DN
QST, January 1984, pages 13-16
For the VIC-20 computer.

"**Build the Billboard Keyboard Keyer**," William B. Jones, KD7S
73 Amateur Radio, July 1983, pages 44-46

"**Keys, Keyers and Keyboards**," Bruce Hale, KB1MW
QST, December 1989, pages 47-49
A general discussion of straight keys, bugs, keyers, iambic keyers, memory keyers and keyboards.

CW FILTERS

"**SuperSCAF and Son — A Pair of Switched-Capacitor Audio Filters**," Rich Arndt, WB4TLM and Joe Fikes, KB4KVE
QST, April 1986, pages 13-19

"**A Tunable CW Filter**," Richard A. Nelson, WBØIKN
QST, October 1983, pages 14-16
Cascaded band-pass filters with voltage controlled center frequencies give this CW filter high selectivity and low ringing.

"**Razor-Sharp CW**," John L. Rehak, N6HI
73 Amateur Radio, February 1986, pages 10-12
An 80 Hz wide CW filter.

"**The DXer's SCF**," Paul Selwa, NB9K
73 Amateur Radio, October 1986, pages 46-52
A switched-capacitor filter for CW and SSB.

"**How to Build a Switched Capacitor Bandpass CW Filter**," Barry L. Ives, AI2T
CQ, January 1986, pages 44-48

"**How to Build a CW Filter for the Novice Operator — Part 1,**" Ed Wetherhold, W3NQN
CQ, February 1985, pages 70-73
An inexpensive, passive inductance-capacitor (LC) filter.

"**How to Build a CW Filter for the Novice Operator — Part 2,**" Ed Wetherhold, W3NQN
CQ, March 1985, pages 72-76

"**A Digital Audio Filter for CW and RTTY,**" Don Kadish, W1OER
Ham Radio, August 1983, pages 60-63
A filter built around the National MF10 integrated circuit.

"**75-Hz Wide Audio Filter,**" Albert A. Roehm, W2OBJ
Ham Radio, January 1990, pages 14-22

"**Switchable Bandwidth Crystal Filter,**" John Pivnichny, N2DCH
Ham Radio, February 1990, pages 22-29
A filter with two different bandwidths, one for CW and one for RTTY.

"**A Deep Notch Resonant Filter,**" Douglas A. Kohl, WØTHM
Ham Radio, February 1990, pages 53-56
Using a 2N2222 transistor, this filter provides a 30 dB notch.

"**Experimenter's Workshop: Build a Crystal Filter,**" Ike Kerschner
Monitoring Times, February 1988, page 80
A crystal filter of several hundred hertz bandwidth for use with receivers having a 455 kHz IF.

"**How to Build the 'Synthetic' Crystal Filter (and Get 'Real' Results),**" Paul D. Carr, N4PC
CQ, April 1990, pages 18-21
A unit with a response similar to that of a crystal filter. It has two CW bandwidths, two phone bandwidths and two notch filters.

"**Passive Audio Filter Design — Part 1: Development and Analysis,**" Stefan Niewiadomski
Ham Radio, September 1985, pages 17-30

"**Passive Audio Filter Design — Part 2: Highpass and Bandpass Filters,**" Stefan Niewiadomski
Ham Radio, October 1985, pages 41-50

"**Narrow-Minded Filtering,**" C. Steer, WA3IAC
73 Amateur Radio, December 1986, pages 48-49
300 Hz, switched-capacitor bandpass filter with narrow passband at −40 dB.

"**Designing and Building Simple Crystal Filters,**" Wes Hayward, W7ZOI
QST, July 1987, pages 24-29
Useful circuits for a variety of crystal filters.

"**Variable-Notch Filters for Receivers**" (A *Hints and Kinks* item), Tom Desaulniers, K4VIZ
QST, January 1985, pages 39-40
A variable-notch filter built around a quad op amp.

"**A Narrow IF Filter for the Heath HW-9 Transceiver**" (A *Hints and Kinks* item), Thomas Niedermeyer, NK6E
QST, June 1990, pages 40-41

"**TS-440S Selectivity Modification for CW**" (A *Hints and Kinks* item), Jeff Elson, KRØO
QST, August 1990, page 37
A remedy for inconveniences experienced when using accessory crystal filters with this rig.

RESONANT CW SPEAKERS

"**An Electro-Acoustic CW Filter,**" J. B. Heaton, G8JFY and R. V. Heaton, G3JIS
QST, April 1983, pages 35-36
Acoustic CW filter built around a drinking glass.

"**Notes on Resonant Speakers for Enhanced CW Reception**" (A *Hints and Kinks* item), Jim Weiss, W9ZMV
QST, January 1989, page 37
Comments and hints on the use of resonant speakers.

"**The One-Note Pipe Organ,**" Glenn Jacobs, WB7CMZ
73 Amateur Radio, August 1979, pages 120-122
A mechanical audio enhancer that produces a clean, single-tone CW note.

"**The Weekender: Solo-16 Acoustic CW Speaker,**" Rick Littlefield, K1BQT
Ham Radio, March 1989, pages 12-14
Clever acoustic CW speaker built around a plastic drink cup.

"**A Resonant Speaker for CW**" (A *Hints and Kinks* item), Wally Millard, K4JVT
QST, December, 1987, page 43
750 Hz resonant speaker made from 2-inch-OD PVC pipe.

CODE OSCILLATORS

"**KOKO — The Kids' Own Kode Oscillator,**" Si Dunn, K5JRN
73 Amateur Radio, February 1987, pages 48-49
A simple code practice oscillator.

"**Code Practice: Have You Seen the Light?**" Homer L. Davidson
73 Amateur Radio, December 1983, pages 10-12
A solar code practice oscillator.

"**A Code Practice Oscillator**," John Pivnichny, N2DCH
Ham Radio, October 1988, pages 65-66
A low-distortion, pure sine wave code practice oscillator made with a single IC.

"**The Hardware-Store Special Code-Practice Set**" (A *Hints and Kinks* item), Art Zavarella, W1KK
QST, January 1990, pages 45-46
A simple, inexpensive, home-brew code-practice set.

CODE TRAINERS

"**You Can Build This Code Trainer**," Harry L. Latterman, K7ZOV
73 Amateur Radio, July 1983, pages 12-16
CPU-controlled Morse trainer that doubles as an electronic keyer.

"**Build the Morsemaster II**," Mike Huddleston, KJ4LN
QST, February 1987, pages 33-38
The Morsemaster II is a code trainer.

"**Processor for Code Tapes**," Andy S. Griffith, W4ULD
Ham Radio, September 1988, pages 61-63
A device to improve the tone and overall audio quality of code tapes.

CW MONITORS

"**How to Build a Super Simple Keying Monitor**," Jim Burtoft, KC3HW
CQ, January 1985, page 39
Three-component keying monitor using a piezoelectric buzzer.

"**Using the Heath HR-1680 Receiver as a CW Monitor**" (A *Hints and Kinks* item), Terry L. Lyon, KA3GCQ
QST, October 1988, page 39

AUTOMATIC CODE READERS

"**The CW Stationmaster**," L.B. Cebik, W4RNL
73 Amateur Radio, January 1984, pages 46-55
A code-copying CW aid.

"**Build This Morse/RTTY Detector — Part 1**," Larry Ashworth, KA7AFR
Radio Electronics, April 1990, pages 33-38
A Morse/RTTY decoder for displaying code on a home computer.

"**Build This Morse/RTTY Detector — Part 2**," Larry Ashworth, KA7AFR
Radio Electronics, May 1990, pages 49-52, 88

"**Robo-Copy**," Michael Hansen, WB9DYI
73 Amateur Radio, October 1990, pages 28-30, 51
Computer/receiver interface unit and computer program for copying Morse code.

CW TRANSMITTERS AND RECEIVERS

"**A Three-Channel CW Emergency Transceiver — Part 1**," Doug DeMaw, W1FB
QST, October 1988, pages 26-30
A one-watt 40-meter CW transceiver.

"**A Three-Channel CW Emergency Transceiver — Part 2**," Doug DeMaw, W1FB
QST, November 1988, pages 17-20

"**A CW Transmitter for 902 MHz**," Douglas L. Hilliard, WØPW
QST, March 1986, pages 32-39

"**The QRP Three-Bander**," Zack Lau, KH6CP
QST, October 1989, pages 25-30
A CW transceiver for 18, 21 and 24 MHz.

"**A Simple 80-Meter Converter**," Doug DeMaw, W1FB
QST, March 1989, pages 18-20
A converter that receives signals between 3.5 and 3.6 MHz in the CW portion of the 80-meter band. It is designed for use with a modified AM transistor radio such as that described in the February 1989 issue of *QST* on pages 33 to 36.

"**A One-Stage 80 Meter CW Transmitter**," Mark A. Boucher, WB3ELL
73 Amateur Radio, February 1989, pages 28-31

"**QRP CW Transceiver**," Bruce Auld, NZ5G and Bill Heishman, N5HNN
73 Amateur Radio, June 1989, pages 20-25, 49
A QRP rig for 30 and 40 meters.

"**CW Transceiver for 20 Meters**," Frank Lee, G3YCC
73 Amateur Radio, June 1989, pages 65-68

"**Army Surplus CW**," Gary Cain, W8MFL and Del Thomas, WL7AKZ
73 Amateur Radio, May 1986, pages 36-37
Conversion tips for the Special Forces PRC-64 transceiver.

"**Flavorig!**" Michael Jay Geier, KB1UM
73 Amateur Radio, November 1989, pages 12-13, 88
An 80-meter CW transceiver built from a Radio Shack transistor radio.

"**Construct Your Own 80 Meter QRP/QSK CW Transceiver**," Salvatore J. Defrancesco, K1RGO
CQ, June 1983, pages 13-18
Multi-featured, VFO-controlled, 1 watt rig for 80.

"**The 30 Meter Fun Machine, a Superhet QRP Transceiver Construction Project — Part 1**," Paul D. Carr, N4PC
CQ, November 1989, pages 32-35

"The 30 Meter Fun Machine, a Superhet QRP Transceiver Construction Project — Part 2," Paul D. Carr, N4PC
CQ, December 1989, pages 32-36

"The 30 Meter Fun Machine, a Superhet QRP Transceiver Construction Project — Conclusion," Paul D. Carr, N4PC
CQ, January 1990, pages 28-31

"A Compact 20-Meter CW Transceiver," Rick Littlefield, K1BQT
Ham Radio, June 1987, pages 8-26
15-watt, 20 meter CW rig in a 1.5- × 4.5- × 6-inch package.

"An NE602-Based QRP Transceiver for 20-Meter CW," Rick Littlefield, K1BQT
Ham Radio, January 1989, pages 9-18

"Cassette Box Special," Michael Jay Geier, KB1UM
73 Amateur Radio, April 1990, pages 46-50
A 5-watt, 80-meter CW transceiver.

"QRP-TX," PA3AUK
Morsum Magnificat, Winter 1987 (no. 6), page 32
300-milliwatt, 40-meter CW transmitter that uses only six electronic components.

"Cubic Incher Mods + 30 Meters," Doug DeMaw, W1FB
QST, September 1990, pages 28-29, 35
Modifications to the 40-meter Cubic Incher QRP transmitter described in July 1982 *QST* that allow operation on 30 meters.

"Some QRP Transmitter Design Tips," Doug DeMaw, W1FB
QST, February 1988, pages 30-32
Tips for improving your CW QRP signal.

"A QRP Transmitter for 30 Meters" (A *Hints and Kinks* item), Frank Pitman, WD4DSS
QST, November 1986, page 43
A 1-watt, 12-volt dc rig for 30 meters.

"Build a QRP Transmitter for 40 meters," Doug DeMaw, W1FB
Monitoring Times, October 1990, pages 92-93
50-milliwatt, crystal-controlled QRP rig for 40 meters.

"How About Used Gear for 80- and 40-Meter CW?" Larry Lisle, K9KZT
QST, March 1987, pages 47-50
Tips on selecting used CW gear for Novice operation.

"A Two-Transistor Transmitter for 30 meters" (A *Hints and Kinks* item), Paul Hoffman, KB4PY
QST, February 1984, pages 46-47
1.5-watt, 30-meter CW rig.

"**The G3IGU Transceiver**," Keith Coates, G3IGU
73 Amateur Radio, June 1989, pages 26-29
A 1-watt, 80-meter CW transceiver.

"**CW Transceiver for 20 meters**," Frank Lee, G3YCC
73 Amateur Radio, June 1989, pages 65-66
VFO-controlled, 20-meter QRP rig

"**The QRP-15 CW Transceiver**," Rick Littlefield, K1BQT
CQ, September 1990, pages 43-49
A full-featured, 5-watt CW transceiver for 15 meters.

"**MOuSe — FeET**," Bill Heishman, N5HNN
73 Amateur Radio, November 1990, pages 36-40
40-watt QSK, CW amplifier.

"**The Fire-Ball QRP Rig**," Bill Brown, WB8ELK
73 Amateur Radio, November 1990, pages 18-21
50-milliwatt, 10-meter CW rig.

"**One-Tube QRP Transmitter**," M.D. Allen
73 Amateur Radio, November 1990, pages 26-29
One-watt, 80- or 40-meter QRP rig built around a single 70L7 tube.

"**TTL Transceiver for 40 Meters**," Rick Lucas, WBØNQM
73 Amateur Radio, November 1990, page 30
350 milliwatt, IC transceiver for 40 meters.

TRANSMITTER KEYING CIRCUITS

"**A CW Keying Interface**," F.A. Bartlett, W6OWP
QST, April 1987, pages 51-53
A device to eliminate start-up distortion in semi-QSK and add adjustable keying weight.

"**Try This Versatile CW Shaper**," Eric P. Nichols, KL7AJ
QST, December 1984, pages 29-30
A circuit that enables you to shape your rig's CW output waveform to your taste.

"**How to Add CW Break-in to the Kenwood TR-9000**,"
John L. Rehak, N6HI
CQ, December 1984, pages 54-55

"**A Band-Aid for Keyer Polarity**," Ed Solov, K2SE
CQ, February 1983, page 89
A device that enables a keyer to be used with rigs having different polarity of keying lines.

"**QSK Amplifier Keying**" (A *Hints and Kinks* item), David J. Rodman, MD, KN2M
QST, August 1986, page 37
Simple method that allows amplifier switching to be done from the exciter without pauses for lost code elements.

"**TS-520 Waveform Shaper**" (A *Hints and Kinks* item), Bruce Cope, KB3LF
QST, April 1985, page 49
Way to soften keying on the TS-520S and TS-520SE transceivers.

"**CW Wave-Shaping Circuit for the TS-820S**" (A *Hints and Kinks* item), Patrick Buller, W7RQT
QST, January 1984, page 46

"**Chirpless Break-In Keying**" (A *Hints and Kinks* item), Berj N. Ensanian, KI3U
QST, August 1985, page 42
Simple technique to cure chirp problems during break-in operation.

"**Reducing Key Clicks**" (A *Hints and Kinks* item), H.H. Hunter, W8TYX
QST, October 1988, page 39
Way to reduce key clicks by adjustment of transmitter's ALC control and DRIVE control.

MISCELLANEOUS CONSTRUCTION PROJECTS

"**A High-Performance CW Demodulator**," Joe Evans, AA4AB
QST, April 1985, pages 23-27
Computer/transceiver interface unit specifically designed for high-quality Morse code reception.

"**A Dichotic Detector for CW**," Douglas A. Kohl, WØTHM
QST, April 1983, pages 32-34
An unusual enhancement aid operating on "psychoacoustic" principles.

"**Micro Morse**," Ralph E. Taggart, WB8DQT
73 Amateur Radio, February 1987, pages 40-43
A CW interface between your rig and your computer.

"**The ROM-less, RAM-less CQ Sender**," G. Gururaj, VU2VIZ
73 Amateur Radio, July 1983, pages 90-93
An automatic code generator.

"**The Threshold Gate — A CW Operator's Accessory**," Fred Brown, W6HPH
CQ, June 1983, page 54
This is used between a receiver and speaker in order to provide a substantial improvement in signal-to-noise ratio.

"**Advanced CW Processor**," Don E. Hildreth, W6NRW
Ham Radio, December 1986, pages 25-29
This enhances CW reception using limiter/filter building blocks.

"**A Carrier-Activated CW Reception Limiter**," Don E. Hildreth, W6NRW
Ham Radio, September 1985, pages 113-120

"**The Basics of Computer CW**," David Oliver, W9ODK
73 Amateur Radio, April 1986, pages 59-61
An unusual circuit that will speed up processing of a BASIC program that copies Morse code.

"**Sending CW: A Digital Approach**," Jonathan Titus, KA4QVK
Ham Radio, June 1983, pages 75-80
This article details the use of digital circuits to accurately time and filter outgoing Morse sent with a straight key or other mechanical key so that dots and dashes are reproducible and well-timed.

"**Morse Time Synthesis**," Lawrence G. Souder, N3SE
Ham Radio, April 1983, pages 16-22
A clock that produces an audible readout of the time in Morse code.

"**Add CW and SSB to any Shortwave Receiver**," Michael A. Covington, N4TMI
Popular Electronics, May 1990, pages 41-42, 108
Very simple device to allow CW reception on shortwave radios that are not so equipped.

"**An Audio-Tape Transmitter Keyer**" (A *Hints and Kinks* item), Arthur C. Erdman, W8VWX
QST, September 1989, page 38
Circuit that enables an audio cassette player with a pre-recorded Morse message to key your transmitter.

"**A Fast-Attack AGC System for QSK CW**" (*A Hints and Kinks* item), Wes Hayward, W7ZOI
QST, September 1990, pages 36-37

"**A Mechanical Morse ID Generator**" (*A Hints and Kinks* item), Alex Comfort, MD, KA6UXR
QST, October 1986, page 48
A Morse ID generator built from a circular printed circuit board with contacts for Morse pulses etched along its circumference.

"**VIC-20 CW Transmitter Interface**" (A *Hints and Kinks* item), W.E. "Bud" Dion, N1BBH
QST, September 1984, page 39
Interface circuit to key transmitter from the sound generated by a Morse code program run on a VIC-20 computer.

"**An Aid to Computer CW**" (A *Hints and Kinks* item), Russ Rennaker, W9CRC
QST, September 1987, page 41
A cure for some of the problems encountered when using a computer/transceiver interface unit for sending CW.

"**Sight and Sound CW**," Don E. Hildreth, W6NRW
Ham Radio, January 1988, pages 10-13
An LED readout circuit that visually aids the operator in picking out a single CW signal from a crowded band.

COMPUTER PROGRAMS FOR CW USE

"**A CW-Program Cartridge for the Atari Computer**," Stephen Stuntz, NØBF
QST, August 1986, pages 34-36
How to modify an existing ATARI program cartridge to contain a CW send/receive program.

"**A CW Keyboard Program for Atari Computers**," Stephen Stuntz, NØBF
QST, February 1985, pages 32-33

"**A CW Receive Program for Atari Computers**," Stephen Stuntz, NØBF
QST, November 1985, pages 51-53

"**A Complete Morse Code System for the VIC-20 Computer**," Barry King, KA7SPU
QST, October 1984, pages 11-16

"**Designing Narrow Band-Pass Filters with a BASIC Program**," William E. Sabin, WØIYH
QST, May 1983, pages 23-29

"**Curious About Code?**" Cass R. Lewart
73 Amateur Radio, February 1986, pages 60-61
A BASIC program for learning or improving your code speed on the TRS-80 computer.

"**Get With the Program and Speed Up Your Code**," Jeff Schwartz, KA2QOU
73 Amateur Radio, April 1986, pages 54-55
A BASIC program to generate random code on the Apple computer.

"**Apple, Morse, and You**," Robert Swirsky, AF2M
73 Amateur Radio, July 1983, pages 54-55
A BASIC CW keyboard program for the Apple II computer.

"**Micro McElroy**," Mike de la Dette, VK3BHM
73 Amateur Radio, July 1983, pages 56-57
A BASIC program for CW reception.

"**Visi Code: The VIC-20 Way to Extra Class**," H. R. Goodsell, W7LTH
73 Amateur Radio, August 1983, pages 14-15
A computer program for learning the Morse code on the VIC-20 computer.

"**A Computer Program that Acts as a Straight-Key Keyer**," Phil Anderson, WØXI
CQ, February 1983, pages 54-55
A program for the Apple computer that improves straight-key sending.

"**Morse Code Teaching Tools for the C-64 and 128**," Dennis L. Green, KB8CS
Ham Radio, January 1988, pages 35-36
A method of controlling between-character timing for increased proficiency and greater speed.

"**Morse Code Computer Tutor**," Lawrence G. Souder, N3SE
Ham Radio, June 1985, pages 45-46
A computer program for learning Morse code on the VIC-20.

"**Semi-Random Code Practice — A New Approach for Thirteen**," Larry W. Loen, NØHCQ
CQ, July 1988, pages 52-53
A BASIC program that generates semi-random code to aid in breaking the 13-WPM code barrier.

"**Experimenter's Workshop: Learn Morse Code on Your TRS-80**," Howard A. Chorost, KC7AC
Monitoring Times, January 1985, pages 34-35
A BASIC program to generate random code practice.

OF HISTORICAL INTEREST

"**Phillips Who?**" Raymond B. Brightman, WA6HDX
QST, August 1971, pages 74-75
Story of the inventor of the Phillips Code.

"**The Phillips Code — Part 1**," Bill Welsh, W6DDB
CQ, November 1979, pages 56-65

"**The Phillips Code — Part 2**," Bill Welsh, W6DDB
CQ, December 1979, pages 35-37

"**Meet the Champion**," Tony Smith, G4FAI
Morsum Magnificat, Summer 1989 (no. 12), pages 24-26
The story of Harry A. Turner, the world record holder for high-speed sending by straight key.

"**Harold S. Bride, Heroic Telegrapher of the R.M.S. Titanic**," David O. Norris, N8HKV
CQ, February 1989, pages 36-38

"**Ice Ahead!**" Tony Smith, G4FAI
Morsum Magnificat, Spring 1987 (no. 3), pages 1-7
Story of the *Titanic* disaster and the role Morse telegraphy played in the event.

"**Drifting Along the Telegraph Trail — Part 1**," Dr. William C. Hess, W6CK
73 Amateur Radio, February 1988, pages 34-35
Memoirs of a former telegrapher.

"**Drifting Along the Telegraph Trail — Part 2**," Dr. William C. Hess, W6CK
73 Amateur Radio, March 1988, page 50

"**Drifting Along the Telegraph Trail — Part 3**," Dr. William C. Hess, W6CK
73 Amateur Radio, April 1988, pages 50-51

"**Drifting Along the Telegraph Trail — Part 4**," Dr. William C. Hess, W6CK
73 Amateur Radio, June 1988, page 38

"**Marconi Station Reborn on Cape Cod**," Stephen C. Place, WB1EYI and Peter O'Dell, N1UM
QST, May 1978, pages 42-43

"**Old CC Calling**," Everett L. Slosman
Monitoring Times, February 1989, pages 10-12
Story of the site of the first U.S. transatlantic wireless telegraph station, its history and status today as a national park.

"**Marconi — DXer or Con-Man?**" A. D. Taylor, G8PG
Morsum Magnificat, Summer 1987 (no. 4), pages 44-47
Analysis of the controversy that sometimes surrounds Marconi's reception of the signals from Poldhu, England in 1901.

"**The First Radio-Telegraph Transmission**," Rowland F. Pocock
Morsum Magnificat, Winter 1989 (no. 14), pages 1-6

"**The Coming of CW**"
QST, March 1964, pages 71-74
Story of the transition from spark gap to CW.

"**Sunken Key**," Tony Smith, G4FAI
Morsum Magnificat, Summer 1987 (no. 4), pages 31-33
An interesting story about a telegraph key linked with the sinking of the German High Seas Fleet at Scapa Flow in 1919.

"**Demise of the Press Sounder**," Fred Barnes, G4LDE
Morsum Magnificat, Autumn 1989 (no. 13), pages 26-28

"**The Heliograph**," Louis Meulstee, PAØPCR
Morsum Magnificat, Summer 1989 (no. 12), pages 1-7

"**Solar Powered Telegraphy**," Don de Neuf, WA1SPM
Morsum Magnificat, Summer 1987 (no. 4), pages 14-16
Story of the heliograph.

"**Earth Current Telegraphy**," Louise Meulstee, PAØPCR
Morsum Magnificat, Autumn 1988 (no. 9), pages 1-8
An excellent article on the early development of single wire telegraph systems using the earth as a return circuit.

"**The Heritage of Telegraphy**," Frederick T. Andrews
IEEE Communications Magazine, August 1989 (vol. 27, no. 8), pages 12-18
The story of telegraphy in the 1800's.

"Morse's First Message Travels Coast-to-Coast — This Time by Satellite" (News item)
IEEE Spectrum, August 1974, page 24

"Early Days: Celerity of Morse's Telegraphy"
Morsum Magnificat, Autumn 1986 (no. 1), pages 15-17
An excerpt from *The Electric Telegraph* (1854) describing the talents of telegraphers of the day as they listened to Morse being "clicked" out while it was being recorded on a strip of moving paper.

"Tom Edison — Telegrapher," Tony Smith, G4FAI
Morsum Magnificat, Autumn 1986 (no. 1), pages 24-32
Story of Edison's involvement with the telegraph and the impact it had upon his life.

"Frederick George Creed — Inventor Extraordinary," Fred Barnes, G4LDE
Morsum Magnificat, Autumn 1990 (no. 17), pages 22-25
Story of the inventor of the Creed High Speed Automatic Printing Telegraph System, used for newspaper communications in England during the early 1900's.

MISCELLANEOUS

"In the Blink of an Eye," Ronald J. Curt, N6VQQ
QST, July 1990, page 44
Fascinating story of a bed-ridden patient, unable to speak, whose only means of communication was Morse code transmitted by blinking his eyes.

"Morse Code from the Heart," Donna Burch, W8QOY
QST, July 1990, page 45
Story of a disabled patient who communicated with his wife by tapping out Morse code with his fingers.

"Samuel F. B. Morse is Alive and Well, Thank You," Samuel F. B. Morse III, W6FZZ
Morsum Magnificat, Summer 1987 (no. 4), pages 34-38
Interesting story of the great grandson of Samuel F. B. Morse.

"The CW Hotshots," George Hart, W1NJM
QST, July 1967, pages 68-71, 148
Discussion of what makes a good CW operator, especially in the traffic-handling community.

"High Speed Hand Sending," Bill Dunbar, AD9E
Morsum Magnificat, Spring 1989 (no. 11), pages 18-20
A discussion of high speed hand sending records and a challenge offered by the British Broadcasting Corporation's program "Record Breakers" to anyone who wishes to attempt to break the existing record.

“**Using Your Transceiver’s Notch Filter as a Zero-Beat Indicator**” (A *Hints and Kinks* item), Brice Wightman, VE3EDR
QST, July 1988, page 45
Another way to zero-beat CW signals.

“**Super-CW — Potentially Error-Free Morse Code**,” Phil Anderson, WØXI
CQ, March 1983, pages 58-61

“**Coherent CW — Amateur Radio’s New State of the Art?**” Raymond C. Petit, W7GHM
QST, September 1975, pages 26-27

“**Coherent CW — The Concept**,” Charles Woodson, W6NEY
QST, May 1981, pages 11-14

“**Coherent CW — The Practical Aspects**,” Charles Woodson, W6NEY
QST, June 1981, pages 18-23

“**The Metamorphosis of CQ**,” Norris K. Maxwell, K5BA
CQ, March 1978, page 27
Insight and information on this often-used abbreviation.

“**First-Class CW Operators’ Club**,” Tony Smith, G4FAI
Morsum Magnificat, Spring 1988 (no. 7), pages 4-7
The story behind this elite organization.

“**HSC, VHSC, SHSC, EHSC**”
Morsum Magnificat, Autumn 1986 (no. 1), pages 7-8
Story behind the High Speed Club (25 WPM requirement), Very High Speed Club (40 WPM requirement), Super High Speed Club (50 WPM requirement) and Extremely High Speed Club (60 WPM requirement).

“**The Radio Spectrum: A Gift to the Weatherwise, Part 6 — Decoding Marine Weather**,” Bert Hunealt
Monitoring Times, June 1984, pages 5, 28-29
How to understand coded CW marine weather broadcasts.

“**Morse Marked Money**,” Don K. de Neuf
Morsum Magnificat, Summer 1987 (no. 4), page 8
Brief but very interesting story on the only known coin (Canadian) inscribed with Morse dits and dahs, spelling out “WE WIN WHEN WE WORK WILLINGLY.”

NOTE: *Morsum Magnificat* is published in England. It is *the* magazine for Morse telegraphy. Despite the fact that it is not readily obtainable in libraries, most back issues may be secured by writing to the address shown in the *periodicals* section of the appendix.

Pg A 8

Author's Note

For years there has been a debate within the amateur community as to whether a Morse Code sending and receiving test should be part of the requirement for an Amateur Radio license. Proponents for a test give arguments that range from allegations that it will help to keep "riffraff" out of the hobby, to claims that it is essential in some emergency situations. Opponents offer rebuttals that newer communication modes make the code outdated, or that those who favor such a test want it only because "they had to do it, so therefore others should also." These arguments are just a few simplified examples of hundreds of arguments that have been presented for or against this issue.

As a result of this ongoing debate the ARRL appointed a special committee to study the implications of a no-code amateur license. The results of the committee's findings were presented to the ARRL Executive Committee during its April 1, 1989 meeting. After studying these findings, the ARRL Board of Directors authorized the filing of a petition for rule making with the Federal Communications Commission for a new class of Amateur Radio license that would not require a Morse Code test. This action represents the culmination of a long standing dispute.

The FCC responded favorably to the petition, but in a modified form. The eventual outcome was a codeless license that is almost identical to the old Technician class. It allows Technician class operation on all amateur frequencies above 30 MHz and it requires the same written test as the original Technician license, but it has no Morse Code requirement.

Despite the fact that this new license is now a reality, nothing said within this book will change. And now the focus of all amateurs should not be on what has transpired in the past, but on what will happen in the future. We should welcome these new licensees into our ranks regardless of our feelings on the code/no-code issue, and offer them the same friendly assistance, cooperation and consideration that our Amateur's Code requires us to show all newcomers to our great hobby.

Appendix

Abbreviations

A	Ampere
ac	Alternating current
AF	Audio frequency
AFRCC	Air Force Rescue Coordination Center
AGC	Automatic gain control
AM	Amplitude modulation
ANL	Automatic noise limiter
ANT	Antenna
ARQ CW	Automatic Repetition reQuest CW
ARRL	American Radio Relay League
ARRS	Aerospace Rescue and Recovery Service
AVC	Automatic volume control
BC	Broadcast
BFO	Beat-frequency oscillator
BUG	Semiautomatic telegraph key
CAL	Calibrate
CST	Central Standard Time
CCW	Coherent CW
CD	Civil Defense
CW	Continuous wave
dB	Decibel
dc	Direct current
DST	Daylight Savings Time
DX	Distance
ELT	Emergency Locator Transmitter
EME	Earth-moon-earth
EPIRB	Emergency Position-Indicating Radio Beacon
EST	Eastern Standard Time
FAA	Federal Aviation Agency
FCC	Federal Communications Commission
FEC CW	Forward Error Correction CW
FM	Frequency modulation
GHz	Gigahertz
GMT	Greenwich Mean Time
GND	Ground
HF	High Frequency
Hz	Hertz
kHz	Kilohertz
LCD	Liquid crystal display
LED	Light-emitting diode
LF	Low Frequency
LID	Poor sender
LSB	Lower sideband
MAYDAY	Voice distress signal
MF	Medium Frequency
MHz	Megahertz
MST	Mountain Standard Time

NASA	National Aeronautics and Space Administration
NOAA	National Oceanic and Atmospheric Administration
PST	Pacific Standard Time
QCWA	Quarter Century Wireless Association
RCC	Rescue Coordination Center
RF	Radio frequency
RIT	Receiver incremental tuning
RST	Readability-Strength-Tone
RTTY	Radioteletype
RX	Receive
SAR	Search and rescue
SARSAT	Search and Rescue Satellite-Aided Tracking
SHF	Super-high Frequency
S/N	Signal-to-noise
SOS	Morse code distress signal
SSB	Single sideband
SS CW	Standard Speed CW
SWL	Shortwave listener
T/R	Transmit/Receive
TV	Television
TX	Transmit
UHF	Ultrahigh Frequency
USB	Upper sideband
UTC	Coordinated Universal Time
VHF	Very-High Frequency
VLF	Very-Low Frequency
WPM	Words per minute
XCVR	Transceiver
XTAL	Crystal
Z	Zulu (same as UTC)

Code Speed Requirements for Amateur and Commercial Licenses

Amateur Licenses

Novice	5 WPM
Technician	None (5 WPM to obtain HF privileges)
General	13 WPM
Advanced	13 WPM
Extra	20 WPM

Commercial Licenses

Third class	16-WPM cipher groups and 20-WPM plain text
Second class	16-WPM cipher groups and 20-WPM plain text
First class	20-WPM cipher groups and 25-WPM plain text

UTC Time

UTC stands for Coordinated Universal Time. It was once called GMT (Greenwich Mean Time) and it is used to eliminate the confusion resulting from different time zones around the word. UTC does away with other items of confusion, including date changes when crossing the International

Date Line, standard time versus daylight savings time, and use of the same hour in the AM or PM (3 AM and 3 PM, for example). You should be familiar with UTC because it is standard in Amateur Radio activities.

UTC is the time at the 0° meridian (which happens to run through Greenwich, England). Time zones around the world are set up so that local times change by one hour for every 15° change in longitude. As a result, Eastern Standard Time (EST) is five hours earlier than UTC, Central Standard Time (CST) is six hours earlier, Mountain Standard Time (MST) is seven hours earlier and Pacific Standard Time (PST) is eight hours earlier.

UTC also makes use of the military format when expressing hours. That is, hours past 12:00 noon do not revert to 1:00 again, but rather continue on, with 1 PM being 1300 hours. 11 PM is 2300 hours, and midnight is 2400 hours (or 0000 hours of the following day). Speaking of days, be sure not to lose a day in conversion. For example, if your local time and date is 2100 EST December 15, then UTC time and date would be 0200 December 16. All confusion can be eliminated if you purchase a 24-hour clock; set it permanently to UTC and leave it.

When looking at a time written in UTC, you may find it noted in one of three ways. 11 PM, for example, might be written as 2300 UTC, 2300 GMT or 2300Z (for "Zulu").

As a note of interest, UTC is determined from the average time of 18 atomic clocks positioned around the globe. These clocks operate on a principle that determines time intervals by measuring the frequency of light given off by atoms. They are so accurate that some will gain or lose only a few seconds every 100,000 years.

Code Practice Schedules

A number of groups offer on-the-air code practice. Most of these are Amateur Radio organizations.

From ARRL HQ in Newington, Connecticut, W1AW transmits a regular schedule of code practice runs as listed on the following page. Transmissions are made simultaneously on eight different frequencies. This schedule is subject to change. The current schedule appears in *QST* magazine and is also available from ARRL HQ for an SASE.

In addition to code-practice transmissions, the ARRL provides code-proficiency certificates for those who qualify. Special certificate-qualifying runs are made twice per month from ARRL HQ. A certificate is issued for demonstration of code-copying ability at 10 WPM. Endorsement stickers are then provided for those able to copy 15, 20, 25, 30, 35 and 40 WPM. To earn a certificate or endorsement sticker, you must copy at least one consecutive minute of perfect code out of a five-minute transmission at any given speed. Contact ARRL HQ (enclose an SASE) for

W1AW schedule

Pacific	Mtn	Cent	East	Sun	Mon	Tue	Wed	Thu	Fri	Sat
6 am	7 am	8 am	9 am			Fast Code	Slow Code	Fast Code	Slow Code	
7 am	8 am	9 am	10 am			Code Bulletin				
8 am	9 am	10 am	11 am			Teleprinter Bulletin				
9 am	10 am	11 am	noon							
10 am	11 am	noon	1 pm		**Visiting Operator Time**					
11 am	noon	1 pm	2 pm							
noon	1 pm	2 pm	3 pm							
1 pm	2 pm	3 pm	4 pm	Slow Code	Fast Code	Slow Code	Fast Code	Slow Code	Fast Code	Slow Code
2 pm	3 pm	4 pm	5 pm	Code Bulletin						
3 pm	4 pm	5 pm	6 pm	Teleprinter Bulletin						
4 pm	5 pm	6 pm	7 pm	Fast Code	Slow Code	Fast Code	Slow Code	Fast Code	Slow Code	Fast Code
5 pm	6 pm	7 pm	8 pm	Code Bulletin						
6 pm	7 pm	8 pm	9 pm	Teleprinter Bulletin						
6^{45} pm	7^{45} pm	8^{45} pm	9^{45} pm	Voice Bulletin						
7 pm	8 pm	9 pm	10 pm	Slow Code	Fast Code	Slow Code	Fast Code	Slow Code	Fast Code	Slow Code
8 pm	9 pm	10 pm	11 pm	Code Bulletin						
9 pm	10 pm	11 pm	Mdnte	Teleprinter Bulletin						
9^{45} pm	10^{45} pm	11^{45} pm	12^{45} am	Voice Bulletin						

W1AW's schedule is at the same local time throughout the year. The schedule according to your local time will change if your local time does not have seasonal adjustments that are made at the same time as North American time changes between standard time and daylight time. From the first Sunday in April to the last Sunday in October, UTC = Eastern Time + 4 hours. For the rest of the year, UTC = Eastern Time + 5 hours.

❑ ***Morse code transmissions:***

Frequencies are 1.818, 3.5815, 7.0475, 14.0475, 18.0975, 21.0675, 28.0675 and 147.555 MHz.

Slow Code = practice sent at 5, 7½, 10, 13 and 15 wpm.

Fast Code = practice sent at 35, 30, 25, 20, 15, 13 and 10 wpm.

Code practice text is from the pages of *QST*. The source is given at the beginning of each practice session and alternate speeds within each session. For example, "Text is from July 1992 *QST*, pages 9 and 81" indicates that the plain text is from the article on page 9 and mixed number/letter groups are from page 81.

Code bulletins are sent at 18 wpm.

❑ ***Teleprinter transmissions:***

Frequencies are 3.625, 7.095, 14.095, 18.1025, 21.095, 28.095 and 147.555 MHz.

Bulletins are sent at 45.45-baud Baudot and 100-baud AMTOR, FEC Mode B. 110-baud ASCII will be sent only as time allows.

On Tuesdays and Saturdays at 6:30 PM Eastern Time, Keplerian elements for many amateur satellites are sent on the regular teleprinter frequencies.

❑ ***Voice transmissions:***

Frequencies are 1.855, 3.99, 7.29, 14.29, 18.16, 21.39, 28.59 and 147.555 MHz.

❑ ***Miscellanea:***

On Fridays, UTC, a DX bulletin replaces the regular bulletins.

W1AW is open to visitors during normal operating hours: from 1 PM until 1 AM on Mondays, 9 AM until 1 AM Tuesday through Friday, from 1 PM to 1 AM on Saturdays, and from 3:30 PM to 1 AM on Sundays. FCC licensed amateurs may operate the station from 1 to 4 PM Monday through Saturday. Be sure to bring your current FCC amateur license or a photocopy.

In a communication emergency, monitor W1AW for special bulletins as follows: voice on the hour, teleprinter at 15 minutes past the hour, and CW on the half hour.

Headquarters and W1AW are closed on New Year's Day, President's Day, Good Friday, Memorial Day, Independence Day, Labor Day, Thanksgiving and the following Friday, and Christmas Day.

more information about these runs.

Another amateur station, W1NJM, also provides on-the-air code practice. These sessions are geared toward the high-speed operator, with runs of 20 to 70 WPM being provided. Currently, sessions are begun at 0130 UTC Monday on 7023 kHz and at 0130 UTC Thursday on 3523 or 14023 kHz (depending upon propagation conditions). Listen to the regular weekly transmissions for announcements of speeds and sequences, as well as future practice schedules.

RST System

The RST system is a method of noting the quality of a received signal. This quality is indicated by a three-digit number, the first digit of which indicates readability (R), the second digit of which indicates strength (S) and the third digit of which indicates tone (T). For example, you may hear one station say to another "Your RST 478." By referring to the chart below, you will see that the station's signal is readable with practically no difficulty, is moderately strong and has a near perfect tone with a slight trace of modulation.

Readability (R)

1—Unreadable.
2—Barely readable, occasional words distinguishable.
3—Readable with considerable difficulty.
4—Readable with practically no difficulty.
5—Perfectly readable.

Signal strength (S)

1—Faint signal, barely perceptible.
2—Very weak signal.
3—Weak signal.
4—Fair signal.
5—Fairly good signal.
6—Good signal.
7—Moderately strong signal.
8—Strong signal.
9—Extremely strong signal.

Tone (T)

1—Extremely rough ac tone, and very broad.
2—Very rough ac tone, and broad.
3—Rough ac tone, rectified but not filtered.
4—Rough ac tone with some trace of filtering.
5—Filtered and rectified ac tone, but strong ripple modulation.
6—Filtered tone with definite trace of ripple modulation.
7—Almost perfect tone but trace of ripple modulation.
8—Near perfect tone with slight trace of ripple modulation.
9—Pure note.

Amateur Operating Privileges

160 METERS

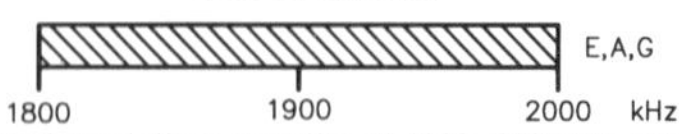

Amateur stations operating at 1900–2000 kHz must not cause interference to the radiolocation service and are afforded no protection from radiolocation operations.

80 METERS

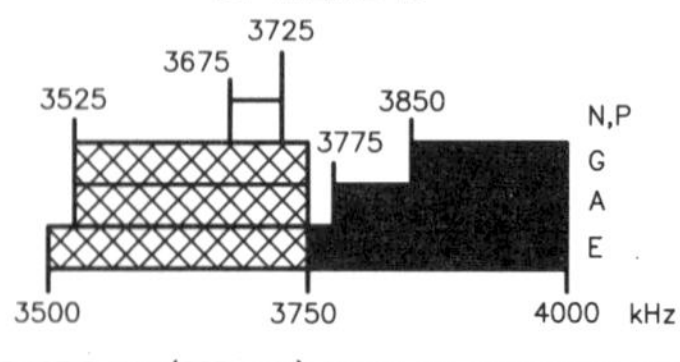

5,167.5 kHz (SSB only) Alaska emergency use only.

40 METERS

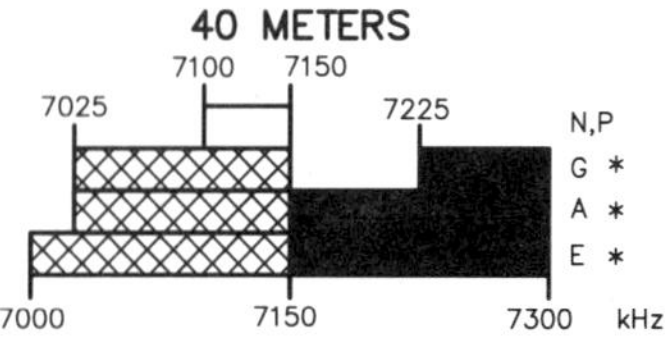

* Phone operation is allowed on 7075–7100 kHz in Puerto Rico; US Virgin Islands and areas of the Caribbean south of 20 degrees north latitude: and in Hawaii and areas near ITU Region 3, including Alaska.

30 METERS

Maximum power on 30 meters is 200 watts PEP output. Amateurs must avoid interference to the fixed service outside the US.

20 METERS

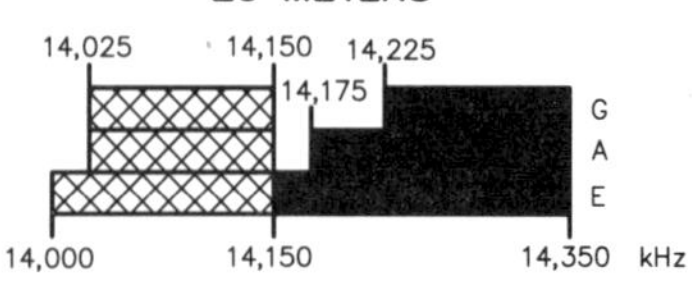

17 METERS

15 METERS

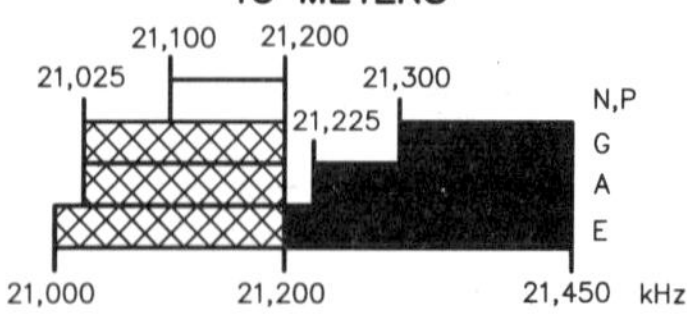

KEY

= CW, RTTY and data

= CW, RTTY, data, MCW, test, phone and image

= CW, phone and image

= CW, RTTY, data, phone and image

= CW and SSB

= CW only

E = AMATEUR EXTRA
A = ADVANCED
G = GENERAL
P = TECHNICIAN PLUS
T = TECHNICIAN
N = NOVICE

** Geographical and power restrictions apply to these bands. See The FCC Rule Book for more information about your area.

12 METERS

10 METERS

28,100 28,500

N,P

E,A,G

28,000 28,300 29,700 kHz

Novices and Technicians are limited to 200 watts PEP output on 10 meters.

6 METERS

50.1

E,A,G,P,T

50.0 54.0 MHz

2 METERS

144.1

E,A,G,P,T

144.0 148.0 MHz

1.25 METERS

E,A,G,P,T,N

222.0 225.0 MHz

Novices are limited to 25 watts PEP output on 1.25 meters.

70 CENTIMETERS **

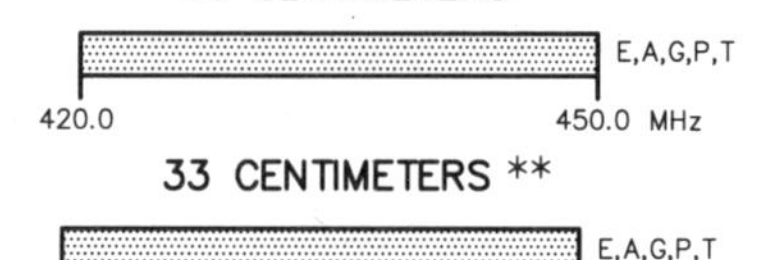

23 CENTIMETERS **

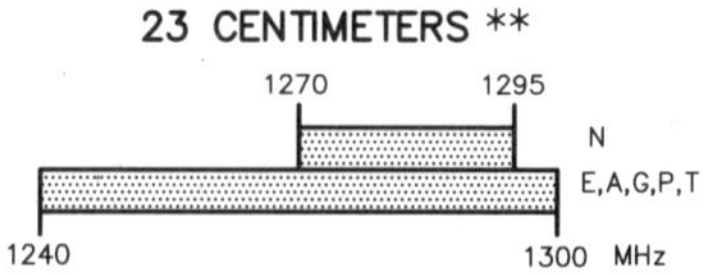

Novices are limited to 5 watts PEP output from 1270 to 1295 MHz.

Above 23 Centimeters:
All licensees except Novices are authorized all modes on the following frequencies:
2300–2310 MHz, 2390–2450 MHz, 3300–3500 MHz, 5650–5925 MHz, 10.0–10.5 GHz, 24.0–24.25 GHz, 47.0–47.2 GHz, 75.5–81.0 GHz, 119.98–120.02 GHz, 142–149 GHz, 241–250 GHz, All Above 300 GHz

Books and Reference Manuals

The ARRL Handbook for Radio Amateurs
American Radio Relay League, 225 Main St, Newington, CT 06111

This is the standard technical reference work for all Amateur Radio operators. It was first published in 1925 and more than five million copies have been sold since then. The complete range of Amateur Radio is covered in depth, including electronic theory, digital electronics and communications, frequency synthesizers, RF power-amplifier design, transceivers, troubleshooting, putting a station together, monitoring and direction-finding and much more. Many construction projects are included.

Now You're Talking!
American Radio Relay League, 225 Main St, Newington, CT 06111

This book is an excellent study guide for either the Novice or the code-free Technician class license. With clear explanations and complete question pools, *Now You're Talking* provides everything needed to pass either examination. In addition, the book offers comprehensive information that will allow the new ham to get on the air quickly and easily.

The ARRL Operating Manual
American Radio Relay League, 225 Main St, Newington, CT 06111

This book is the source for contemporary, how-to-do-it Amateur Radio operating techniques. In addition to particulars on proper CW operating, this lavishly illustrated book covers everything from DXing to digital, from shortwave listening to satellites. The *Operating Manual* also features nuts-and-bolts advice on contesting, traffic handling, DXing, net controlling, talking to OSCAR, gearing up for RTTY, international frequency allocations, international voltage/current requirements, and much more.

The Code Book
by Robert J. Halprin, K1XA
Tiara Publications, PO Box 493, Lake Geneva, WI 53147.

This is a comprehensive book on the code, covering operating techniques, keys, keyers, contesting, DXing, traffic handling, QRP operating and more.

Vintage Radio
by Morgan E. McMahon
Vintage Radio, Box 2045, Palos Verdes Peninsula, CA 90274

This is an excellent book for those interested in antique radios. It is filled with pictures of old wireless gear, crystal sets, telegraph keys, spark gaps and more. Equipment shown covers the era from the beginnings of radio to about 1929.

Periodicals

Morsum Magnificat
G. C. Arnold Partners, 9 Wetherby Close, Broadstone,
Dorset BH18 8JB, England

An excellent quarterly booklet of about 50 pages per issue, containing an amazing assortment of information on telegraphy and related subjects. Originally published in Holland beginning in 1983, the first English- language edition appeared in 1986.

This publication "provides international in-depth coverage of all aspects of Morse telegraphy, from its earliest concept to the present time. It is for all Morse enthusiasts, amateur or professional, active or retired. It brings together material which would otherwise be lost to posterity, providing an invaluable source of interest, reference and record relating to the traditions and practice of Morse."

A **highly recommended** periodical and a must for the serious Morse enthusiast. It is easy to subscribe to in the states by surface or airmail.

Dots and Dashes
Morse Telegraph Club, Inc, 1101 Maplewood Drive, Normal, IL 61761

A fine quarterly newspaper for those interested in telegraphic history. This journal of the Morse Telegraph Club, Inc, contains many articles, mostly of an historical nature, all submitted by members.

The Old Timer's Bulletin
The Antique Wireless Assn, 59 Main St, Bloomfield, NY 14469-9336

A quarterly bulletin of the Antique Wireless Association. In its 25 to 50 pages each quarter, this publication discusses club activities, restoration of equipment, vacuum-tube collecting, and more.

QST
American Radio Relay League, 225 Main St, Newington, CT 06111-1494

This is the official monthly magazine of the American Radio Relay League, devoted entirely to Amateur Radio, and is an interesting, lively way to keep on top of everything that is happening in ham radio. Every facet of ham radio is covered, including receiving and transmitting equipmeet, antennas, FCC rules, basics for beginners, contests, hamfests, propagation, consumer-oriented product reviews and ARRL lab analysis of new products. More pictures, more drawings, more how-to articles—written by hams for hams. *QST* is read by more amateurs than all other ham magazines combined!

Organizations and Associations

American Radio Relay League
225 Main St, Newington, CT 06111-1494

The ARRL is the official organization for Amateur Radio operators the world over. Founded in 1914 "of, by and for the radio amateur," ARRL is the leader in preserving and expanding amateur frequencies, advancing technology and furnishing each member with a strong voice in Amateur Radio and League affairs. Through ARRL, hams have a strong, effective voice with regulatory agencies and government bodies nationally and internationally. ARRL fights successfully for radio amateurs' interests on all fronts. And in addition to publishing *QST* magazine, the League's famous library of amateur publications, headed by the annual *ARRL Handbook for Radio Amateurs*, the all-time best-selling technical publication, is recognized worldwide for its authoritative leadership. The ARRL HQ staff handles thousands of letters and countless telephone calls from all over to help individual members with technical, operating, legal and regulatory questions, and sponsors a myriad of programs and activities for radio amateurs. W1AW, the voice of ARRL HQ, transmits daily code practice sessions and regular bulletins for the benefit of all amateurs and would-be amateurs. If you are not yet a radio amateur, but are ready to take that first step, remember that ARRL stands ready to help you along the way. Contact the Educational Activities Department at ARRL HQ for information, including referral to a local ARRL-affiliated Amateur Radio Club near you (800-32-NEW-HAM).

Antique Wireless Assn, Inc
59 Main St, Bloomfield, NY 14469-9336

The AWA is an international, historical radio organization whose primary purpose is the exploration and preservation of the history of wireless communication. It sponsors regional and national meetings and exhibits, publishes or assists in publishing historical work, documents the history of radio by preparing slides and movies, and distributes a quarterly journal, *The Old Timer's Bulletin*. AWA also maintains a museum in Bloomfield, New York.

Morse Telegraph Club, Inc
1101 Maplewood Drive, Normal, IL 61761

This organization was formed for the purpose of perpetuating the art of telegraphy and to keep alive the memories of those who were in the forefront of its development. The club was organized in 1942 and now has 49 chapters in the United States and Canada. Anyone with a knowledge of the code is allowed to join. On the last Saturday in April, all chapters meet to commemorate the birthday of Samuel F. B. Morse (April 27, 1791). At that time, the Western Union Telegraph Company in the United States and Canadian Telecommunications in Canada jointly provide a telegraph circuit connecting all chapters. A quarterly newspaper, *Dots and Dashes*, is sent to all members.

National Science Center for Communications
1 7th St, Augusta, GA 30901

TheNational Science Center for Communications is a joint venture of the Federal Government and a not-for-profit private foundation which supports the operation. When completed, this center will include a 250 seat interactive theater, display modules, classrooms and a main exhibit hall. Ground breaking for the complex took place in November, 1988. Completion of various stages of the project is due over a period of years, with *Fort Discovery,* an interactive, hands-on exhibit building scheduled to open in early 1997. The purpose of the NSCC is to provide a resource center of educational activities and materials to enhance and support, nationwide, a higher quality of science and mathematics education while improving the training and understanding of communications and electronics.

Quarter Century Wireless Assn
159 East 16th St, Eugene, OR 97401-4017

The QCWA is an organization of Amateur Radio operators who have been licensed for 25 years or more. It has over 15,000 members in more than 130 chapters worldwide. Meetings are scheduled on a local and national basis and a quarterly newsletter is distributed.

Society of Wireless Pioneers
PO Box 530, Santa Rosa, CA 95402

Organized to collect, research and record the history of communications. Each member must have been, at one time, a professional radio operator, and must have held a government license. Associate membership is available to those who do not meet the above requirements but have been engaged in some aspect of the communications field and exhibit more than a cursory interest in the association's goals. The society publishes a quarterly newsletter, *The World Wireless Beacon.*

US Army Signal Corps Assn
PO Box 7740, Fort Gordon, GA 30905

Comprised of men and women who now serve, or who have previously served, in the US Army Signal Corps in either a military or civilian capacity. It maintains a museum, provides audio and visual aids of historical interest, has a recognition program for outstanding graduates of the Signal Corps and keeps its membership up to date on the latest advances in communications technology through its journal, the *Army Communicator.*

Veteran Wireless Operators Association
Office of the Secretary, 46 Murdock Street, Fords, NJ 08863

The object and general purpose of the Veteran Wireless Operators Association is "to foster and extend an esprit de corps among wireless operators, recognize meritorious service, and to acquaint the general public of such works." Full membership is extended only to those who have held a first or second class radiotelegraph or radiotelephone license and served for at least three years as an operator of a radiotelegraph or radiotelephone station. Associate membership may be obtained at the discretion of the Membership

Committee if it is deemed that such membership would be beneficial to the organization. The VWOA publishes a quarterly newsletter and a Yearbook.

First-Class CW Operators' Club
C. J. Page, G4BUE, Secretary
Alamosa, The Paddocks, Upper Beeding, Steyning,
West Sussex, BN44 3JW, ENGLAND

Founded in 1938, the FOC is a difficult club to join because its membership is limited to a total of 500. Openings do occur from time to time, however. To be accepted, an amateur must be capable of sending and receiving at least 25 WPM and must be nominated by five existing members without his or her knowledge. The club basically stands for excellence in CW operating practices, setting standards for others to emulate.

European CW Association
Oscar Verbanck, ON5ME, Pylyserlaan 58,
B-8458 Oostduinkerke, Belgium

This is an association of 14 individual European Amateur Radio clubs, comprising a total membership of about 10,000 CW operators. Its principal purpose is "to support and encourage Amateur Radio CW operating." Clubs included in the group are listed below.

Club	*Acronym*	*Country Where Headquartered*
The G-QRP Club	G-QRP	United Kingdom
Scandinavian CW Activity Group	SCAG	Scandinavia
German CW Activity Group	AGCW-DL	Germany
TOPS CW Club	TOPS	United Kingdom
High Speed Club	HSC-DARC	Germany
Very High Speed Club	VHSC	Holland
Super High Speed Club	SHSC	Belgium
FISTS Morse Club	FISTS	United Kingdom
Benelux QRP Club	B-QRP	Belgium /Holland/ Luxembourg
Italian Naval Old Rhythmers Club	INORC	Italy
Hispania CW Club	HCC	Spain
Belgian Telegraphy Club	BTC	Belgium
Union Francaise des Telegraphistes	UFT	France
First Class C.W. Operators' Club	FOC	United Kingdom

Museums

American Radio Relay League Museum of Amateur Radio
225 Main St, Newington, CT 06111

The ARRL collection, housed in the HQ lobby, consists of about 1400 square feet of shelf area containing antique Amateur Radio equipment. Many pieces of equipment, dating from the invention of wireless and the founding of the League, form the core of this exhibit.

Antique Wireless Assn Radio and Communications Museum
Village Green, Rts 5 & 20, Bloomfield, NY 14469-9336

Although a private museum of the AWA, this collection is open to the public from May to October on Sundays and Wednesdays. It contains old radios, telegraphs, phonographs, televisions and several working, old-time amateur stations.

New England Wireless and Steam Museum, Inc
Tillinghast Rd, East Greenwich, RI 02818

This museum is open Sundays during the summer months and at other times by previous arrangements. School groups, professional associations and clubs are welcome. It contains a large assortment of equipment that traces history from the pre-telegraph age up to the first televisions. Included are crystal sets, spark transmitters, keys, microphones and even a complete replica of a 1920 shipboard radio room. A library of engineering history contains many volumes of interest to telegraphy buffs.

US Army Communications-Electronics Museum
Kaplan Hall, Bldg 275, Avenue of Memories, Fort Monmouth, NJ 07703

This museum is open to public Monday through Friday. On its floors are examples of Army equipment and documents dating from preelectronic communication times up to the period of communication by satellite and laser beams. Included are an early Spanish-American War telegraph, Vietnam-era radios, telegraphs from World War I, an Apollo display and signal equipment from the Civil War.

Young-Morse Historic Site
370 South Rd (Rte 9), Poughkeepsie, NY 12602

This was the home of Samuel F. B. Morse and is now a National Historic Landmark. The principal attractions here are the 100-acre estate, 24-room house and a large collection of early-American furnishings. The Morse Room houses cabinets containing telegraph equipment.

Historic Speedwell
333 Speedwell Avenue, Morristown, NJ 07960

This national historic site preserves the Vail Homestead Farm where Alfred Vail worked with Morse to improve versions of the early telegraph. On display are documents, models and instruments illustrating the development of this device.

Sources of Telegraph Equipment

The following pieces of equipment are produced by the listed manufacturers.

Audio Filters

Autek
Datong Electronics
MFJ Enterprises

Automatic CW Readers

Microcraft
Universal Electronics

Code Practice Oscillators

AMECO Publishing
MFJ Engerprises
William M. Nye
Radio Shack

Code-Practice Cassette Tapes

AMECO Publishing
ARRL
Gordon West Radio School, Inc.
Wrightapes

Code-Practice Computer Software

AEA
ARRL
Gordon West Radio School
Renaissance Development
W5YI Group

Computer Software for Station Operation

Electrosoft
Fundamental Services
Microsystems Software
Personal Database Applications

CW Interface for Home Computer

Advanced Electronic Applications
Crown Microproducts
HAL Communications
Kantronics
MFJ Enterprises

CW Keyboards

MFJ Enterprises
Microcraft

Electronic Keyers

Advanced Electronic Applications
Autek Research
MFJ Enterprises
MSC
William M. Nye
Palomar Engineers
Ten-Tec

Random-Code Generators

Advanced Electronic Applications
Datong Electronics
MFJ Enterprises

Receivers

Drake
ICOM America
Kenwood
Ten-Tec
Radio Shack
Yaesu Electronics

Telegraph Keys

Bencher
William M. Nye
Vibroplex

Addresses of Manufacturers

Advanced Electronic Applications, Inc
PO Box 2160
Lynnwood, WA 98036

AMECO Publishing Corp
275 Hillside Ave,
Williston Park, NY 11596

Autek Research
PO Box 8772
Madeira Beach, FL 33738

Bencher, Inc
831 North Central Ave
Wood Dale, IL 60191

Datong Electronics
c/o Gilfer Associates
PO Box 239
52 Park Avenue
Park Ridge, NJ 07656

Drake
PO Box 3006
Miamisburg, OH 45343

HAL Communications Corp
Box 365
Urbana, IL 61801

Electrosoft
PO Box 1462
Loveland, CO 80539

ICOM America, Inc
2380-116th Avenue NE
Bellevue, WA 98004

Kantronics
1202 East 23rd St
Lawrence, KS 66046

Kenwood USA Corp
2201 E Dominquez St
PO Box 22745
Long Beach, CA 90801-5745

MFJ Enterprises, Inc
Box 494
Mississippi State, MS 39762

Microcraft Corp
PO Box 513Q
Thiensville, WI 53092

Microsystems Software, Inc
600 Worcester Road — Suite B2
Framingham, MA 01701-5360

William M. Nye Company, Inc
PO Box 1877
Priest River, ID 83856

Palomar Engineers
Box 462222
Escondido, CA 92046

Personal Database Applications
1323 Center Dr
Auburn, GA 30203-3318

Radio Shack
Tandy Corporation
1 Tandy Center—Suite 300
Fort Worth, TX 76102

Renaissance Development
Box 640
Kilen, AL 35645

Ten-Tec
1185 Dolly Parton Hwy
Sevierville, TN 37862

Universal Radio, Inc
6830 Americana Parkway
Reynoldsburg, OH 43068

The Vibroplex Co, Inc
11 Midtown Park East
Mobile, AL 36606-4141

W5YI Group
PO Box 565101
Dallas, TX 75356

Gordon West Radio School, Inc
PO Box 2013
Lakewood, NJ 08701

Wrightapes
235 E Jackson S-1
Lansing, MI 48906

Yaesu Electronics Corp
17201 Edwards Road
Cerritos, CA 90701

INDEX

ARRL MEMBERS

This proof of purchase may be used as a $.80 credit on your next ARRL purchase. Limit one coupon per new membership, renewal or publication ordered from ARRL Headquarters. No other coupon may be used with this coupon. Validate by entering your membership number from your *QST* label below:

MORSE CODE BOOK

PROOF OF PURCHASE

FEEDBACK

Please use this form to give us your comments on this book and what you'd like to see in future editions, or e-mail us at **pubsfdbk@arrl.org** (publications feedback).

Where did you purchase this book?
☐ From ARRL directly ☐ From an ARRL dealer

Is there a dealer who carries ARRL publications within:
☐ 5 miles ☐ 15 miles ☐ 30 miles of your location? ☐ Not sure.

License class:
☐ Novice ☐ Technician ☐ Technician Plus ☐ General
☐ Advanced ☐ Amateur Extra

Name ______________________ ARRL member? ☐ Yes ☐ No

Call Sign ______________________

Daytime Phone ( ) ______________ Age ______________

Address ______________________________

City, State/Province, ZIP/Postal Code ______________________

If licensed, how long? ______________________________

Other hobbies ______________________

____

Occupation ______________________

For ARRL use only	MC : TEL
Edition	2 3 4 5 6 7 8 9 10 11 12
Printing	3 4 5 6 7 8 9 10 11 12

From ______________________

____

____

Please affix postage. Post Office will not deliver without postage.

EDITOR, MORSE CODE: THE ESSENTIAL LANGUAGE
AMERICAN RADIO RELAY LEAGUE
225 MAIN STREET
NEWINGTON CT 06111-1494

· · · · · · · · · · · · · · · · · · · please fold and tape · · · · · · · · · · · · · · · · · · ·